THE SCIENCE DELUSION

Copyright

The Science Delusion

National Library of Australia Cataloguing-in-Publication entry

Author: Fred Smith

Title: The Science Delusion

ISBN: 978-0-9923128-2-4

Target Audience: Academics, scientists, evolutionists, creationists, Christians.

Religion and science.
Science--Social aspects.
Astronomy.

Dewey Number: 201.65

Published with the assistance of:www.loveofbooks.com.au

Contents

PREFACE

Spokesmen for evolution use Karl Popper's philosophy of science to argue that 'creationism is not science'.

Sir Karl Raimund Popper (1902-1994) claimed that his testability criterion solved *the problem of demarcation* of science. Popper's testability criterion defines three major categories for theories —'tested successfully' (corroborated), 'tested but failed' and 'not testable'.

Regarding the 'not testable' category Popper said that if we cannot test a conjecture, then the conjecture is 'not science', and spokesmen for evolution use this understanding against creationism. Spokesmen for evolution use Popper's philosophy to permanently classify creationism as 'not-science' without looking at the evidence. For example, the National Academy of Sciences (of USA) writes on page 25 of their *Science and Creationism* booklet: "Creationism, intelligent design, and other claims of supernatural intervention in the origin of life or of species are not science because they are not testable by the methods of science."

Spokesmen for evolution need to say something categorical about creationism because they need to keep religion out of the science classrooms. Spokesmen for evolution sometimes say that 'creationism is religion not science', whatever that means. And spokesmen for evolution sometimes say that evolution is 'not anti-religion' but that it is 'only anti-fundamentalism'.

'Only anti-fundamentalism' is an effective thing to say because the 'fuzzy religion factors' are far more pervasive than the 'creation religion factor'. 'Only anti-fundamentalism' affirms the popular belief in a semi-real god, a smoky heaven, and a fireless hell. Christian fundamentalists believe in a literal creation and they believe in a real God and a real heaven and hell. And I like to think that we Christian fundamentalists have more interest in real and ordinary truth than in effective truth.

Although I try to keep the specifics of my fundamentalism (my real agenda) away from this first book in the *Science Delusion* series, readers need to at least understand that the first words of the Bible say: "In the beginning God created the heavens and the earth." Readers need to have a clear understanding of the fact that we fundamentalists care about the theory of evolution because the theory of evolution insists that 'there is no evidence for a real God who can do things, because he has never done anything'.

The understanding that 'there is no God that can do things because he has never done anything' came from what I like to call the God-is-dead enlightenment. Darwin's science was enlightened because it had a problem with a real God doing things. Newton's science wasn't enlightened because Newton's science did not have a problem with a real God that can do things.

In Part B of this book, I argue that after Newton, science zigged when it should have zagged. However, before we get to that, I need to get you interested in Popper's philosophy of science (Part A). Although Popper did not intend that we use his philosophy to provide a derogatory classification of a theory, he did want to address delusional attitudes to explanatory theories. Popper argued that testing a theory provided a way of escaping this unscientific attitude.

Not only did Popper claim that his testability philosophy provides the solution to the problem of 'demarcation of science from non-science' but he also claimed that it solves the 'problem of induction' and both of these topics are important to *The Science Delusion.* Chapter 1 discusses the history of the problem of induction. Chapter 2 provides a justification for induction. Chapter 3, the final chapter in part A, discusses Popper's thoughts about his solution to the problem of demarcation and the problem of induction.

Part B applies the understandings of Part A to the creation-evolution argument. Claims like 'creationism is not science' are considered along with 'philosophical naturalism' and other issues that come from enlightened science.

I like to take readers back to ancient cultures so that we can understand how it came to be that science has no respect for the idea of a real God who can do things. Richard Dawkins argues against every kind of god, however, Dawkins is not representative of vocal science. Science typically does not have a problem with god doing something billions of years ago. Compared to the God of the Bible, and even the gods of ancient gentile religions, the god of science is a do nothing god. Although there are a few spokesmen for evolution such as Richard Dawkins, who agree with my perspective, many scientists prefer to say things like 'science is anti-fundamentalism not anti-religion'. Many scientists do not want a clear understanding of the fact that modern science only allows a fuzzy do nothing god.

In the interest of whatever progression is possible, I try to take some care with my perspective. I will try to avoid 'my historical facts' by basing my arguments on secular accounts of history and culture. Bertrand Russell's book *History of Western Philosophy* is a source I cite from extensively. Russell's capacity to retain a disinterested approach to Christianity is what makes his book such a great source of information. My disagreements with Russell can often be put down to different priorities or different assumptions — which is to say a different perspective rather than different facts. Something similar can be said of Adrian Desmond & James

Moore's Darwin. I also use David Lindberg's *The Beginnings of Western Science* and Edward Grant's *The Foundations of Modern Science in the Middle Ages* and Steven Shapin's *The Scientific Revolution* to fill in some history. And even authors, such as Richard Dawkins who have long given up the disinterested approach to Christianity, are a great help to my book because in most respects Dawkins prefers ordinary truth to effective truth.

The heroes of this book are Paley, Behe, Johnson and other creationists. The villains are Darwin, the National Academy of Sciences (of America), Richard Dawkins, Christopher Hitchens and other journalists, scientists and philosophers. Where I have used an online source, the source has been included in the bibliography.

INTRODUCTION

In order to understand the scientific revolution and the enlightenment, it helps to start with ancient Greek science and philosophy that preceded the enlightenment. *An understanding of the heavens* was a very important issue to the scientific revolution.

Aristotle (384 ~ 322 BC) said that the heavens were perfect and unchanging unlike the earth. Aristotle thought that the earth was at the centre of the universe with the sun, stars and planets attached to spheres that rotated about the earth as explained by Edward Grant on pages 65 and 66 of *The Foundations of Modern Science in the Middle Ages:*

> On the nature of the empty celestial spaces, however, Aristotle was quite clear: they were filled with invisible, transparent, ethereal spheres that were nested one with another and each of which turned with regular, uniform, motion. Celestial bodies — planets and fixed stars — were somehow embedded in these spheres and carried around by them. Aristotle based his system upon the earlier mathematical systems of concentric spheres devised by Eudoxus of Cnidus and Callippus of Cyzicus in the fourth century B.C. In the latter's scheme, on which Aristotle directly founded his cosmology of concentric spheres, the planet Saturn, for example, was assigned a total of four spheres that were supposed to account for its celestial position. Of these, one was for Saturn's daily motion; one was for its proper motion along the zodiac, or ecliptic; and two represented its observed retrograde motions along the zodiac. Aristotle transformed Callippus's mathematical spheres into a system of real, earth-centered, physical celestial orbs that were collectively coterminous with the

> celestial region. To prevent the transmission of Saturn's zodiacal and retrograde motions to Jupiter, the planet immediately beneath Saturn, Aristotle added three "unrolling," or counteracting, spheres for Saturn. The purpose of the three unrolling spheres was to counteract the motions of three of Saturn's four spheres, with the exception of the sphere representing the daily motion (because the daily motion was common to all planets, each was assigned a special sphere for that purpose, thus acknowledging that the daily motion was transmissible through each set of planetary spheres).

If celestial spheres support the stars and the planets, then it seems natural to reason that an agent such as God must have put the celestial spheres there to support the planets and the stars because the celestial spheres suggest purpose in design and purpose in design suggests an agent. This tells us that Aristotle's science was going to have some affinity with creationist religions such as Christianity, Judaism, and Islam.

The understanding that the earth was fixed at the centre of the universe was perhaps as much due to a common sense perspective as it was to Aristotle's science. When Copernicus (1473 – 1543) proposed a sun-centred universe the Church was initially tolerant of Copernicus' theory. That remained true until Galileo (1564 – 1642) used the newly invented telescope to support Copernicus' sun-centred system and to contradict other aspects of Aristotle's science. Later in his life, Galileo again troubled the Church with his scientific theories and consequently, he was tried by the inquisition, forced to recant of his sun centred beliefs, and put under house arrest.

When I use the word Church with a capital 'C', I mean an old organisation with secular power and an attachment to the New Covenant of the Bible. The King James version of the Bible, and many versions that followed the King James version, use the word 'church' but the original Greek word really means a local 'assembly' of New Covenant believers. King James perhaps wanted to be head of the 'Church of England'. In western Europe from not long after the fall of Rome until 1517, when Luther protested against the Roman Catholic Church using his *The Ninety-Five Theses*, there was no church other than the Roman Catholic Church. Luther's protests established a protestant church before Copernicus' theory was published. Unlike the slow and careful response of the Catholic Church, the protestant response to Copernicus' theory was immediate, hostile and unequivocal as Russell explains on pages 512-513 of his *History of Western Philosophy:*

> Some of the men to whom Copernicus communicated his theory were German Lutherans, but when Luther came to know of it, he was profoundly shocked. 'People give ear,' he said, 'to an upstart astrologer who strove to show that the earth revolves, not the heavens or the firmament, the sun and the moon. Whoever wishes to appear clever must devise some new system, which of all systems is of course the very best. This fool wishes to reverse the entire science of astronomy; but sacred Scripture tells us that Joshua commanded the sun to stand still, and not the earth.' Calvin similarly, demolished Copernicus with the text: 'The world also is stablished, that it cannot be moved' (Psa. xciii. 1), and exclaimed:' 'Who will venture to place the authority of Copernicus above that of the Holy Spirit?' Protestant clergy were at least as bigoted as Catholic ecclesiastics; nevertheless there soon came to be much more liberty of speculation in Protestant than in Catholic countries, because in Protestant countries the clergy had less power. The important aspect of Protestantism was schism, not heresy, for schism led to national Churches, and national Churches were not strong enough to control the lay government. This was wholly a

> gain, for the Churches, everywhere, opposed as long as they could practically every innovation that made for an increase of happiness or knowledge here on earth.

Russell says that Copernicus was a devout Lutheran. If Copernicus was alive today we would say that he was a fundamentalist. Perhaps, though, it will help to relieve some truth-tensions if we fundamentalists acknowledge that Luther's style rather than Copernicus' style better represents modern fundamentalists. This is to say that if a typical modern day fundamentalists were back in Luther's time, then he would be supporting Luther's argument against Copernicus, because that is what we do best.

Kepler perfected the mathematics of the Copernicus model with the use of elliptical orbits. Whereas, Copernicus continued to believe that celestial spheres supported the planets, Johannes Kepler (1571 – 1630) omitted to provide a mechanism to move the planets in elliptical orbits and consequently René Descartes (1596 – 1650) filled that gap by suggesting that a whirlpool of invisible particles impacted on and moved the planets.

Further progress on the modelling the planets had to wait for an understanding of dynamics of bodies and in this case, Aristotle rather than the Church seems to be the main hindrance to science. Aristotle taught natural motion and violent motion where natural motion is expressed as a tendency of a body to move in a certain direction — such as the tendency of a rock to fall or smoke to rise. Aristotle also argued that it is natural for bodies to come to rest. According to Edward Slowik's *Stanford Encyclopaedia* article on Descartes' Physics, Descartes rejected Aristotle's understanding of motion and derived a concept of inertia that is similar to the modern concept of inertia:

> He [Descartes] argues, "because experience seems to have proved it to us on many occasions, we are still inclined to believe that all movements cease by virtue of their own nature, or that bodies have a tendency towards rest. Yet this is assuredly in complete contradiction to the laws of nature; for rest is the opposite of movement, and nothing moves by virtue of its own nature towards its opposite or own destruction" (Pr II 37). While one can find several natural philosophers whose earlier or contemporary work strongly foreshadowed Descartes' achievement in the first and second laws—namely, Galileo and Isaac Beeckman—the precise formulation put forward in the *Principles of Philosophy* is quite unique (especially as regards the second law, since both Galileo and Beeckman appear to sanction a form of circular inertial motion, which possibly betrays the influence of the Scholastic's circular motion of the celestial element). A fascinating blending of Scholasticism and the new physics is also evident in above quotation, since Descartes invokes the logic of contrary properties in his statement that "nothing moves by virtue of its own nature towards its opposite or own destruction". That is, rest and motion are opposite or contrary states, and since opposite states cannot (via the Scholastic principle) transform into one another, it follows that a body at rest will remains at rest and a body in motion will remains in motion. Consequently, Descartes has employed a Scholastic/Medieval argument to ground what is possibly the most important concept in the formation of modern physics, namely inertia.

Descartes (1596 – 1650) dismisses the understanding of rest and motion learned from experience and instead he invokes what seems to be mere words-and-definitions (or what the Stanford Encyclopaedia calls a 'scholastic principle') to derive scientific knowledge. The fact that Descartes' method seems so incomprehensible to everyone on this side of the scientific revolution suggests that the issue of how to derive confident scientific knowledge was another very important issue resolved by the scientific revolution.

The issue of how we derive our scientific knowledge is a part of the overall theory of knowledge — what philosophers call the problem of epistemology. How do we know that 3x5 = 15? How do we know that parallel lines will never meet? Scientific knowledge is about the world we live in. Scientific knowledge is about ‘all mules being sterile’ or the ‘the sun rising each day’. Philosophers wanted to know how to justify a trust in our scientific knowledge. How do we know that ‘all mules are sterile’? How do we know that the sun will rise tomorrow? One school of ancient Greeks argued for rationalism and another argued for empiricism.

Ancient Greek rationalists maintained that people are born with some kind of innate *knowledge* of the world we live in or innate ability to reason about the world we live in just as we can reason that 1+1 =2. Rationalists believed that we can derive the various laws of the universe by thinking about the universe. Rationalists believed that there was such a thing as self-evident empirical truth, as if the real world was like the analytic world of mathematics and logic.

Empiricists such as Aristotle believed that we should learn about the real world by observing the real world. In book 2 section 21 of his *Prior Analytics*, Aristotle writes: “Nothing prevents a man who knows both that A belongs to the whole of B, and that B again belongs to C, thinking that A does not belong to C, e.g. knowing that every mule is sterile and that this is a mule, and thinking that this animal is with foal: for he does not know that A belongs to C, unless he considers the two propositions together. So it is evident that if he knows the one and does not know the other, he will fall into error. And this is the relation of knowledge of the universal to knowledge of the particular.” Here, Aristotle is going from givens or the empirical premises (the idea that all mules are sterile) to particulars (such as a deduction about a specific mule). Aristotle's demonstrative logic is a part of the analytic world of logic whereas his premises that 'all mules are sterile' came from observation of the real world.

Descartes' Scholasticism was a Catholic invention, apparently inspired by ancient Greek rationalism. Perhaps even Descartes would be happy to base the understanding that 'mules are sterile' on observation; however as we have seen when it came to the issue of inertia, Descartes scholasticism assumed that this problem could be solved without the need to conjecture empirical premises. Descartes thought he was deriving his knowledge of the world on the basis of logic and the fundamental truth that 'nothing moves by virtue of its own nature towards its opposite or own destruction'. Descartes' scholasticism seems to assume that the real world was like the analytic worlds of mathematics and logic, except that where mathematicians use symbols, operators and equations, Descartes uses words-and-definitions and logic.

In theory, the Catholic Church should have had a good understanding of the difference between the empirical world and the analytic world because the concept of a creator separates the empirical world from the analytic world. The empirical world is the world made by God. According to Grant's account of the dispute between the Catholic Church and Philosophers cited in the notes, the Catholic Church used the understanding 'that God could do what he liked short of a logical contradiction' against the philosophers. Whatever the sense of Catholic reasoning on earlier occasions, it seems remarkable that only 43 years before Newton published his *Principia*, Descartes' saw fit to use Scholasticism to express the understanding that it is impossible for God to create a universe where things slowed down of their own accord.

God apparently has no choice about the fact that 1 +1 = 2. God has no choice but to agree with Aristotle's logic. Empirical truths are different. In another universe, it may be that only some mules are sterile. Objects need not be conserved if God so desired. Moving objects of their own accord could come to rest if God wanted it that way. Things could fall upwards if God wanted. In other words, God can make apples any size and colour that he wants but he has no choice but to make little green apples little and green because analytic worlds are worlds constructed from definitions.

We know for sure that little green apples are little and green and we must agree with Aristotle's logic given the empirical premises; however, there is still the question of how we can know if the empirical premises are true. How can we know that all mules are sterile? Aristotle argued that we get to know the premises by induction. Induction is a process of generalising our experience beyond the reach of our experience. We can learn by experience that all tested mules are sterile but unless we have been taking a representative random sample of all mules, then we need inductive thinking to say that 'all mules are sterile'. Unless we have been sampling all mules, we need inductive thinking even to say that 'most mules are sterile' or even that 'mules are probably sterile'.

For the sake of word efficiency we could try and say that people who use inductive thinking are empiricists. However, that would mean that everyone who expected the sun to rise was an empiricist because experience is always past tense and the past is not a sample of the future.

If instead of saying that empiricists are people who trust inductive thinking we could try and say that empiricists are those people who do not trust in rationalism, however, we may still find ourselves in a few difficulties. It is said that Aristotle was an empiricist. And if there was such a thing as an ancient Greek empiricist it seems that Aristotle was one.

However, Aristotle and his many disciples that followed him reasoned too much, asserted too much and experimented too little. Aristotle said that heavy objects fall faster than light objects apparently because he did not experiment. Aristotle argued that 'a vacuum could not exist because a vacuum is nothing and nothing cannot exist'. Perhaps it helps to say that Aristotle and his disciples were empiricists who reasoned too much and experimented too little because they lived in a culture of rationalism.

Bertrand Russell argues the ancient Greek thinkers who came before Aristotle were more scientific Greeks and that Aristotle led the western world up the garden path. Aristotle's celestial sphere theory, for example, does not seem very scientific because it suggests that the spheres were put there for a purpose and purpose implies an agent. This kind of thinking is called teleological thinking.

Teleological thinking was not the only garden path. When we compare how Descartes argued with how Aristotle argued, then it seems difficult to escape the impression that rationalism and the failure to experiment was far more problematic to science than teleological thinking. We also know that Aristotle's understanding of physics and astronomy was almost completely wrong. Not that being wrong was necessarily a problem, because science had to start from somewhere, but that Aristotle was completely wrong and that he was held in extremely high regard throughout western culture for such a very long time.

Euclidian geometry has been held in high regard for even longer than Aristotles science and philosophy. Euclidian geometry conjectured five axioms which were supposed to be the basis of all geometry: 'a straight line may be drawn through any two points', 'any terminated straight line may be extended indefinitely', 'a circle may be drawn with any given centre and any given radius', 'all right angles are congruent' and 'for a given line and a point there is only one line parallel to the first line that passes through the point'. From these axioms, Euclidian geometry provided an apparently complete and secure framework for geometry. Euclid's thinking was recognised as one of the all time greatest mathematical achievements, and this became a problem to science because Euclid's thinking was based on fundamental principles that can be recognised by the mind (self-evident truths), rather than truths discovered by experience. Euclid's success conveyed the understanding that we can discover reliable and robust empirical truth by thinking about problems (words and-definitions).

Russell (p67) attributes the first use of rationalism to Parmenides (~515~440 BC) who came hundreds of years before Euclid. Perhaps then we should allow that the success of Pythagoras (~570~440 BC) sent the ancient Greeks up the words-and-definitions garden path. Pythagoras' famous theory has some of the features of Euclidian geometry because triangles exist in the real world, but whatever the case, the astounding success of Euclidian axiomatic thinking seems to support Descartes' words-and-definitions thinking. And given that some of Aristotles Comments on Scientific Knowledge cited in the notes seem more modern than Descartes' words-and-definitions approach, that came 2000 years later, it makes sense to say that too much respect for Euclid's method and too much respect for Aristotles' understanding led western science up the garden path.

As already noted, Galileo supported Copernicus' challenge to Aristotle's understanding of the heavens. On pages 61-69 of his *Dialogue Concerning Two New Sciences* Galileo (1564 – 1642) speaking through Salviati and Sagredo to Simplicio, doubts that Aristotle ever tested his theory that 'heaver objects fall faster than small ones'. Sagredo says: "But I, Simplicio, who have made the test and can assure you that a cannon ball weighing one or two hundred pounds, or even more, will not reach the ground by as much as a span ahead of a musket ball weighing only half a pound, provided both are dropped from a height of 200 cubits".

Not many years after Galileo (1564 – 1642) and Descartes (1596 – 1650), but in a time when Aristotle was no longer held in high regard, Newton (1642-1727) proposed a universal law of gravitation and some laws of dynamics. Rising and falling tides, falling objects, projectiles, orbiting moons, pendulums, planets and stars all obeyed the same simple law of gravity. When combined with the laws of dynamics, the position of all planets and moons could, in theory, be accurately predicted for any future or past time. It seems easy to understand that the step from reasoning that 'nothing moves by virtue of its own nature towards its opposite or own destruction' to being able predict the positions of the planets with a simple universal law that applied to every object in the universe, was a truly astonishing *scientific* enlightenment.

Not long after Newton, David Hume (1711-1776) wrote in his *An Enquiry Concerning Human Understanding* Section XII, Part 3: "If we take in our hand any volume; of divinity or school metaphysics, for instance; let us ask, Does it contain any abstract reasoning concerning quantity or number? No. Does it contain any experimental reasoning, concerning matter of fact and existence? No. Commit it then to flames: for it can contain nothing but sophistry and illusion." Hume seems to take the view that all words-and definition arguments came from a deliberated strategy of rhetoric and intellectualism. Hume apparently owes this great insight to the fact that he was born on this side of the scientific enlightenment.

Even on this side of the scientific enlightenment the division between the analytic world and the empirical world is not as sharp as we might like. We know that God could have made little green apples any size and colour he wanted provided he made them little and green but what about space and time? If God made space, then where was he when he did that? If he made time when did he do that? We also know that before Newton, Galileo the experimental scientist, rationalised (through his character Salviati on pages 61-69 of his *Dialogue Concerning Two New Sciences*) that small stones and large stones fall at the same speed: "But if this is true, and if a large stone moves with a speed of, say, eight while a smaller moves with a speed of four, then when they are united, the system will move with a speed less than eight; but the two stones when tied together make a stone larger than that which before moved with a speed of eight. Hence the heavier body moves with less speed than the lighter; an effect which is contrary to your supposition. Thus you see how, from your assumption that the heavier body moves more rapidly than the lighter one, I infer that the heavier body moves more slowly." This means that there is more than one type of rationalism.

Apparently, there is also more than one type of argument. Had Descartes' argument turned out to contradict experience, then Descartes had the option of changing his words to argue something else, whereas if Galileo's argument turned out to contradict experience, then Galileo would have discovered a paradox. According to Hume's anti-rationalistic philosophy, true paradoxes are impossible in the real world because it is impossible to discover the real world by thinking about the real world.

Hume is better known for his anti-empirical arguments than for his anti-rationalistic thinking. Despite the fact that Newton's science seemed to confirm the empirical method, Hume argued against the empirical method. Hume pointed out that observation and experience only tell us what happened previously and that we need to invoke inductive thinking to apply our past experience to new situations. Hume argued that there was no possibility of justifying inductive thinking with logic. Hume claimed that just because we observe a certain behaviour many times, we have no reason to expect the behaviour to continue. According to Hume's philosophy of knowledge, the fact that all known mules are sterile does not justify the conclusion that some unknown mule outside our sample domain is sterile or even probably sterile. Hume seemed to believe that induction worked (because he used inductive thinking in his arguments) but he did not believe it was possible to justify induction with logic.

Bertrand Russell similarly claimed that the principle of induction cannot be proved by either logic or experience and that science is not possible without induction. However, Popper argued that not only is it not possible to justify the inductive reasoning process, but that such a reasoning process is in fact invalid. Popper claimed that his testability criterion provided a solution to the problem-of-induction and he claimed that we do not learn by induction but we learn by testing and falsification. According to Popper, when we learn that mules A, B and C are sterile, then all we can say from such experience is that 'not all mules are fertile' but of course Aristotle's logic already said as much.

Whatever Hume, Russell, and Popper would tell us about the validity of inductive thinking we can be sure that Hume, Russell, and Popper all used inductive thinking. Even to say that rationalism was inspired by success (the success of Euclidian geometry) is to say that the pre-enlightenment trust in rationalism came from experience.

The National Academy of Sciences (NAS) *Science and Creationism* booklet argues the success of the scientific method. After mentioning the capabilities and uses of the Satellite Navigation System, NAS argue on p viii of the Preface of their *Science and Creationism* booklet that:

> The tremendous success of science in explaining natural phenomena and fostering technological innovation arises from its focus on explanations that can be inferred from confirmable data. Scientists seek to relate one phenomenon to another and to recognize the causes and effects of phenomena. In this way they have developed explanations for the changing of the seasons, the movements of sun and stars, the structure of matter, the shaping of mountains and valleys, the changes in the positions of continents over time, the history of life on Earth, and many other natural occurrences. By the same means, scientists have also deciphered which substances in our environment are harmful to humans and which are not, developed cures for diseases, and generated the knowledge needed to produce innumerable labor-saving devices. The concept of biological evolution is one of the most important ideas ever generated by the application of scientific methods to the natural world.

We could also say that the reliability of the theory of evolution comes from the fact that it is based on the same means, procedures, techniques, people, etc, as the rest of science and that evolution is typical science. In fact, science is so reliable that if we were to try and use Popper's philosophy to say that we cannot trust science, most of us would then use the success of science (and the assumption that science is induction) to condemn such a philosophy.

Another pattern we may notice in the NAS words is the pattern of naturalism — even though in the past we conjectured supernatural theories to explain natural occurrences, we are finding that we never need to go beyond naturalism — or rather we are finding that we live in a natural world. In our natural world, we continue to find that '10 bricks plus 10 bricks equals 20 bricks' and we trust that this will always be the case. In other words, the theory of evolution is typical science and typical science is very reliable and it is based on naturalism. These two very persuasive patterns paint the creationist claims as ad hoc, because creationists accept the conclusions and reliability of science in all areas except for science that happens to conflict with what p25 of the NAS *Science and Creationism* booklet identifies as those "body of beliefs that has its origin in doctrinal material rather than scientific observation, interpretation, and experimentation...". Thus, the NAS booklet contrasts the reliability of science and naturalism against the dogmatism of religious fundamentalism. And NAS believes that a categorisation judgement against creationism is the right way to argue because 'creationism is not science'.

PART A:
Philosophy and Induction

Chapter 1 THE PROBLEM OF INDUCTION

The scientific enlightenment taught us that we should use natural laws to understand the universe and that we cannot learn about these natural laws by thinking about the world. If God can do what he likes in the empirical world, then the only chance we have of learning about these natural laws is by observing what God does in the empirical world. In Book 2, section 19 of his *Posterior Analytics* Aristotle argues that we should use induction to learn about the empirical world:

> 2.19. We have already said that scientific knowledge through demonstration is impossible unless a man knows the primary immediate premises...
>
> So out of sense-perception [in the absence of instrumentation all experience came to us via our senses] comes to be what we call memory, and out of frequently repeated memories of the same thing develops experience; for a number of memories constitute a single experience...
>
> We conclude that these states of knowledge are neither innate in a determinate form, nor developed from other higher states of knowledge, but from sense-perception. It is like a rout in battle stopped by first one man making a stand and then another, until the original formation has been restored. The soul is so constituted as to be capable of this process...
>
> Thus it is clear that we must get to know the primary premises by induction; for the method by which even sense-perception implants the universal is inductive.

What Aristotle says about induction in the above citation seems reasonably clear and quite modern. Although

Aristotle seems confident that inductive thinking works, his explanation for how it works seems more psychological than logical, which is also quite modern. About half a century before Newton published his *Principia*, Francis Bacon (1561-1626) attempted to explain in his *Novum Organum* (published 1620) how to use inductive thinking. In Book 1, clause CV Bacon writes:

> CV. In forming axioms, we must invent a different form of induction from that hitherto in use; not only for the proof and discovery of principles (as they are called), but also of minor, intermediate, and, in short, every kind of axioms. The induction which proceeds by simple enumeration is puerile, leads to uncertain conclusions, and is exposed to danger from one contradictory instance, deciding generally from too small a number of facts, and those only the most obvious. But a really useful induction for the discovery and demonstration of the arts and sciences, should separate nature by proper rejections and exclusions, and then conclude for the affirmative, after collecting a sufficient number of negatives. Now this has not been done, nor even attempted, except perhaps by Plato, who certainly uses this form of induction in some measure, to sift definitions and ideas. But much of what has never yet entered the thoughts of man must necessarily be employed, in order to exhibit a good and legitimate mode of induction or demonstration, so as even to render it essential for us to bestow more pains upon it than have hitherto been bestowed on syllogisms. The assistance of induction is to serve us not only in the discovery of axioms, but also in defining our notions. Much indeed is to be hoped from such an induction as has been described.

If we say that all swans that we have seen so far have been white and we ignore all other information, this would apparently be an example of Bacon's induction by enumeration. When we want to use all available information that might have an influence on the pattern, because we think we have some understanding of the causes and effects, then we could call this *empirical induction.* The car mechanic living in the mechanical age suspects that the spark plugs are the cause of the engine failure so he replaces the spark plugs to see if the problem goes away. The car mechanic does not need Bacon to tell him how to do that.

In the empirical age, scientists know how to use tools like induction (and probability and mathematics) in a practical setting to get a handle on the problems of science even if they do not understand that they are using inductive thinking. Newton was very much aware of the fact that his science relied on inductive thinking. On page 320-321 of Book III of his *Principia*, Newton argues that inductive thinking is the correct way to learn about the real world:

> RULE III
>
> *The qualities of bodies, which admit neither intension nor remission of degrees, and which are found to belong to all bodies within the reach of our experiments, are to be esteemed the universal qualities of all bodies whatsoever.*
>
> That abundance of bodies are hard, we learn by experience; and because the hardness of the whole arises from the hardness of the parts, we therefore justly infer the hardness of the undivided particles not only from the bodies we feel but of all others. That all bodies are impenetrable, we gather not from reason, but from sensation. The bodies we have handled we find impenetrable, and thence conclude impenetrability to be a universal property of all bodies whatsoever. That all bodies are moveable, and endowed with certain powers (which we call the *vires inertiae*) of persevering in their motion, or in their rest, we only infer

from the like properties observed in the bodies which we have seen. The extension, hardness, impenetrability, mobility, and *vis inertiae* of the whole, result from the extension, hardness, impenetrability, mobility and *vires inertiae* of the parts; and thence we conclude the least particles of all bodies to be also all extended, and hard and impenetrable, and moveable, and endowed with their proper *vires inertiae*. And this is the foundation of all philosophy... Lastly, if it universally appears, by experiments and astronomical observations, that all bodies about the earth gravitate towards the earth, and that in proportion to the quantity of matter which they severally contain; that the moon likewise, according to the quantity of its matter, gravitates towards the earth; that, on the other hand, our sea gravitates towards the moon; and all the planets one towards another; and the comets in like manner towards the sun; we must, in consequence of this rule, universally allow that all bodies whatsoever are endowed with a principle of mutual gravitation...

RULE IV

In experimental philosophy we are to look upon propositions collected by generalinduction from phenomena as accurately or very nearly true, notwithstanding any contrary hypotheses that may be imagined, till such time as other phenomena occur, by which they may either be made more accurate, or liable to exceptions.

This rule we must follow, that the argument of induction may not be evaded by hypotheses.

...Hitherto we have explained the phaenomena of the heavens and of our sea by the power of gravity, but have not yet assigned the cause of this power. This is certain, that it must proceed from a cause that penetrates to the very centres of the sun and planets, without suffering least diminution of its force; that operates not according to the quantity of the surfaces of the particles upon which it acts (as mechanical causes used to do), but according to the solid matter which they contain, and propagates its virtue on all sides to immense distances, decreasing always in the duplicate proportion of the distances. Gravitation towards the sun is made up out of the gravitations towards the several particles of which the body of the sun is composed; and in receding from the sun decresses accurately in the duplicate proportion of the distances as far as the orb of Saturn, as evidently appears from the quiescence of the aphelions of the planets; nay, and even unto the remotest aphelions of the comets, if those aphelions are also quiescent. But hitherto I have not been able to discover the cause of those properties of gravity from phaenomena, and I frame no hypotheses; for whatever is not deduced from the phaenomena is to be called an hypothesis; and hypotheses, whether methaphysical or physical, whether of occult qualities or mechanical, have no place in experimental philosophy. In this philosophy particular propositions are inferred from the phaenomena, and afterwards rendered general by induction. Thus it was that the impenetrability, the mobility, and the impulsive force of bodies, and the laws of motion and of gravitation, were discovered. And to us it is enough that gravity does really exist, and act according to the laws which we have explained, and abundantly serves to account for all the motions of the celestial bodies, and of our sea.

Richard Westfall explains on pages 730 & 801 of his *Never At Rest — A Biography of Isaac Newton*, that Newton developed the philosophical aspect of his science in part to defend his theories against criticisms that the belief in attraction between bodies was akin to the abandonment of science and to what Gottfried Leibniz identified as "a return to the occult qualities of scholastic philosophy". Newton's explanation suffered from the fact that he could not explain the cause of gravity, whereas Descartes' whirlpool of invisible particles was a true 'scientific' theory that needed no further explaining.

We see from the above that Newton understood that we can only learn about the empirical world by observing the empirical world and that he understood the fallibility of inductive thinking. Even though Descartes' mechanical force by impact causes seem beyond question, Newton's philosophy recognises that we do not even understand very fundamental things like the impenetrability of bodies. And if we do not understand things like the impenetrability of bodies, then we cannot understand Descartes' force by impact causes.

About half a century after Newton published his first edition of the *Principia*, David Hume (1711-1776) argued against the possibility of a logical justification for inductive reasoning. However unlike many modern arguments against induction, Hume also argued against empirical induction. If there is no such thing as induction when we attempt to use cause and effect reasoning (empirical induction), then of course there cannot be such a thing as induction by enumeration (or what Bacon calls puerile induction) because induction by enumeration is empirical induction when we ignore all potential influences on the pattern. Although the modern problem of induction is about induction by enumeration, Hume's problem of induction was more about induction using cause and effect reasoning.

Even though the modern problem of induction ignores Hume's problem of empirical induction, many people do not seem to realise that Hume's problem of empirical induction has already been addressed. Perhaps this lack of understanding owes itself to the absence of an 'empirical induction' label.

Much of this chapter is about ensuring that readers who are familiar with Hume's arguments have a good understanding of what remains of Hume's problem of induction.

Karl Popper's problem of induction is the modern problem of induction and the modern problem of induction focuses on induction by enumeration and ignores Hume's problem of empirical induction. Readers who are not familiar with Hume's problem of induction and who are not interested in understanding the history of Hume's problem of empirical induction need to at least understand the difference between empirical induction and ordinary induction (or induction by enumeration) in order to understand some arguments in part B. However readers who are not familiar with Hume's problem of induction may find some of the arguments in this chapter easier to understand after they have read chapter 2.

Hume also argued against induction by enumeration or what we might call ordinary inductive thinking. If there is no way to justify a trust in induction, then there is no way to justify a trust in the understanding that the sun will rise tomorrow. Induction teaches us that '10 bricks plus 10 bricks equals 20 bricks' and we all think that such laws are beyond question. Apparently, the reason we think that such laws are beyond question is that we give so much credence to familiarity. We have a great deal of experience with such laws and we have always found such laws to be true. Or, we can say that we have so much familiarity with the pattern that we can predict what will happen and we are never wrong so we have no need for any further understanding. That is, even though Hume, Popper and Russell argued against the possibility of a logical justification for induction, it seems that the extreme confidence we achieve from familiar experience comes from inductive reasoning.

Given that Newton accepts that inductive reasoning is both sensible and fallible, Newton must think that inductive reasoning is in some sense probable. Hume insisted that inductive reasoning was not just fallible but that it was not logically probable. That is Hume insisted that inductive thinking could not be justified as probable thinking. Probable thinking tells us that if we take a random sample of a population then we expect the sample to reflect the population. Such knowledge maybe merely probable but it is logically probable knowledge. Inductive thinking is different. In his *System of Logic (1843)* Book 2, Chapter 1, section 3 John Mill (1806 —1873) explains inductive thinking :

> Of Induction, therefore, we shall say no more at present, than that it at least is, without doubt, a process of real inference. The conclusion in an induction embraces more than is contained in the premises. The principle or law collected from particular instances, the general proposition in which we embody the result of our experience, covers a much larger extent of ground than the individual experiments which form its basis. A principle ascertained by experience, is more than a mere summing up of what has been specifically observed in the individual cases which have been examined; it is a generalization grounded on those cases, and expressive of our belief, that what we there found true is true in an indefinite number of cases which we have not examined, and are never likely to examine. The nature and grounds of this inference, and the conditions necessary to make it legitimate, will be the subject of discussion in the Third Book: but that such inference really takes place is not susceptible of question. In every induction we proceed from truths which we knew, to truths which we did not know; from facts certified by observation, to facts which we have not observed, and even to facts not capable of being now observed; future facts, for example; but which we do not hesitate to believe on the sole evidence of

> the induction itself. Induction, then, is a real process of Reasoning or Inference.

In a loose sense, we could say that probability and logic tells us what we already know, whereas, induction predicts what we have no way of knowing. When we use a random sample taken from one part of the population to represent another part of the population (a non-representative sample), then we are using inductive thinking. And when we use the past to predict a future (which is another kind of non-representative sample), then we are using inductive thinking.

1.1 Hume

All of Hume's arguments discussed in the following have been taken from 'Section IV — Sceptical Doubts Concerning the Operations of the Understanding' of his *An Enquiry Concerning Human Understanding.*

In Part I of Section IV, Hume develops the idea that all knowledge of this world (what he refers to as 'matters of fact') is gained via experience. Hume makes the point that the only way we can gain knowledge from experience is via induction — that is via identification and confirmation of consistent pattern. Hume emphasises that although it is obvious that we only learn by experience in complicated situations, this is also true in very familiar and simple situations:

> ALL the objects of human reason or enquiry may naturally be divided into two kinds, to wit, *Relations of Ideas*, and *Matters of Fact*. Of the first kind are the sciences of Geometry, Algebra, and Arithmetic; and in short, every affirmation which is either intuitively or demonstratively certain. *That the square of the hypothenuse is equal to the square of the two sides*, is a proposition which expresses a relation between these figures. *That three times five is equal to half of thirty*, expresses a relation between these numbers. Propositions of this kind are discoverable by the mere operations of thought, without dependence on what is anywhere existent in the universe. Though there never were a circle or triangle in nature, the truths demonstrated by Euclid would for ever retain their certainty and evidence.
>
> Matters of fact, [empirical knowledge or matters of the universe] which are the second objects of human reason,

are not ascertained in the same manner; nor is our evidence of their truth, however great, of a like nature with the forgoing. The contrary of every matter of fact is still possible: because it can never imply a contradiction, and is conceived by the mind with the same facility and distinctness, as if ever so conformable to reality. *That the sun will not rise tomorrow* is no less intelligible a proposition, and implies no more a contradiction than the affirmation, *that it will rise.* We should in vain, therefore, attempt to demonstrate its falsehood. Were it demonstratively false, it would imply a contradiction, and could never be distinctly conceived by the mind...

This proposition, *that causes and effects are discoverable, not by reason but by experience*, will readily be admitted with regard to such objects, as we remember to have once been altogether unknown to us; since we must be conscious of the utter inability, which we then lay under, of foretelling what would arise from them...

But the same truth may not appear, at first sight, to have the same evidence with regard to events, which have become familiar to us from our first appearance in the world, which bear a close analogy to the whole course of nature, and which are supposed to depend on the simple qualities of objects, without any secret structure or parts. We are apt to imagine that we could discover these effects by the mere operation of our reason, without having inferred that one Billiard-ball would communicate motion to another upon impulse; and that we needed not to have waited for the event, in order to pronounce with certainty concerning it. Such is the influence of custom [familiarity] that, where it is strongest, it not only covers our natural ignorance, but even conceals itself, and seems not to take place, merely because it is found in the highest degree.

But to convince us that all the laws of nature, and all the operations of bodies without exception, are known only by experience, the following reflections may, perhaps, suffice. Were any object presented to us, and were we required to pronounce concerning the effect, which will result from it, without consulting past observation; after what manner, I beseech you, must the mind proceed in this operation? It must invent or imagine some event, which it ascribes to the object as its effect; and it is plain that this invention must be entirely arbitrary. The mind can never possibly find the effect in the supposed cause, by the most accurate scrutiny and examination. For the effect is totally different from the cause, and consequently can never be discovered in it. Motion in the second Billiard-ball is a quite distinct event from motion in the first; nor is there any thing in the one to suggest the smallest hint of the other. A stone or piece of metal raised into the air, and left without any support, immediately falls: But to consider the matter a priori, is there any thing we discover in this situation which can beget the idea of a downward, rather than an upward, or any other motion, in the stone or metal? ...

Hence we may discover the reason why no philosopher, [Hume perhaps had Newton in mind] who is rational and modest, has ever pretended to assign the ultimate cause of any natural operation, or to show distinctly the action of the power, which produces any single effect in the universe. It is confessed, that the utmost effort of human reason is to reduce the principles, productive of natural phenomena, to a greater simplicity, and to resolve the many particular effects into few general causes, by reasonings from analogy, experience, and observation. But as to the causes of these general causes, we should in vain attempt their discovery; nor shall we ever be able to satisfy ourselves, by any particular explication of them.

> These ultimate springs and principles are totally shut up from human curiosity and enquiry...
>
> Nor is geometry, [Newton's *Principia* was essentially a book of geometry] when taken into the assistance of natural philosophy, ever able to remedy this defect, or lead us into the knowledge of ultimate causes, by all that accuracy of reasoning of which it is so justly celebrated. Every part of mixed mathematics proceeds upon the supposition that certain laws are established by nature in her operations; and abstract reasonings are employed, either to assist experience in the discovery of these laws, or to determine their influence in particular instances, where it depends upon any precise degree of distance and quantity. Thus, it is a law of motion, discovered by experience, that the moment or force of any body in motion is in the compound ratio or proportion of its solid contents and its velocity; and consequently, that a force may remove the greatest obstacle or raise the greatest weight, if, by any contrivance or machinery, we can increase velocity of that force, so as to make it an overmatch for its antagonist. Geometry assists us in the application of this law, by giving us the just dimensions of all the parts and figures which can enter into any species of machine; but still the discovery of the law itself is owing merely to experience, and all the abstract reasonings in the world could not lead us one step towards the knowledge of it.

Hume perhaps had Descartes' vortices in mind when he used the example of colliding billiard balls to dispute the idea that body-contact causes were necessarily true causes. It may seem that A causes B but it might only be that A happens to cause B in our circumstances and therefore A is not necessarily a true cause. Hume argued that we could not determine the behaviour of billiard balls without the assistance of experience (or a theory that comes from experience) and therefore Hume concluded that our understanding must come from experience and therefore the causes we see are not necessarily true causes.

Hume's distinction between experience knowledge and certain knowledge needs some adjustment because we can say that a contradictory argument can sometimes be 'conceived in the mind' because it will not seem contradictory if it is sufficiently complex. That we can combine tautologies (using say, Boolean algebra or mathematical algebra) to build complicated arguments to make non-obvious predictions is what makes them useful. And some things that we cannot conceive in the mind might be true. We cannot conceive of non-Euclidian geometry in the mind, but this does not mean that such geometries are not true.

Although Hume compares empirical knowledge (matters of fact) with Euclidian geometry, mathematicians tell us that there are other possible geometries and physicists tell us that Euclidian geometry is not necessarily the true geometry for the universe we live in. However, Hume's argument still works because the failure of Euclidian geometry simply becomes an impressive case of the failure of very confident *knowledge*. The failure of Euclidian geometry says that 'obvious and apparently certain truth is not always true' and, therefore, it makes sense to be sceptical about what seems to be common sense.

In part II of section IV, Hume proceeds on the basis that all our knowledge of this world comes by experience and that we confirm the understanding we get from experience by repetition and consistency. But to understand Hume's reasoning we must allow that we only use observation of a pattern in order to learn. We must keep our thinking abstract: we push a button on an electronic sequence machine and a green light comes on, we push it again and a light comes on again, etc. How can we use the past pattern to predict the future pattern without using an assumption, even though we know that throughout all of human history, the green light has always come on? Hume says that even if we are sure that we understand the situation, our understanding only comes from experience, because there is no way for us to be able to predict empirical things without the benefit of experience. If we think we understand the factors then according to Hume, this is only because our experience confirms that the factors have always been the factors:

> If we be, therefore, engaged by arguments to put trust in past experience, and make it the standard of our future judgment, these arguments must be probable only, or such as regard matter of fact and real existence according to the division above mentioned. But that there is no argument of this kind, must appear, if our explication of that species of reasoning be admitted as solid and satisfactory. We have said that all arguments concerning existence are founded on the relation of cause and effect; that our knowledge of that relation is derived entirely from experience; and that all our experimental conclusions proceed upon the supposition that the future will be conformable to the past. To endeavour, therefore, the proof of this last supposition by probable arguments, or arguments regarding existence, must be evidently going in a circle, and taking that for granted, which is the very point in question.
>
> In reality, all arguments from experience are founded on the similarity which we discover among natural objects, and by which we are induced to expect effects similar to those which we have found to follow from such objects. And though none but a fool or a madman will ever pretend to dispute the authority of experience, or to reject that great guide of human life, it may surely be allowed a philosopher to have so much curiosity at least

as to examine the principle of human nature, which gives this mighty authority to experience, and makes us draw advantage from that similarity which nature has placed among different objects. From causes which appear similar, we expect similar effects. This is the sum of all our experimental conclusions. Now it seems evident that, if this conclusion were formed by reason, it would be as perfect at first, and uponone instance, as after ever so long a course of experience. But the case is far otherwise. Nothing so like as eggs; yet no one, on account of this appearing similarity, expects the same taste and relish in all of them. It is only after a long course of uniform experiments in any kind, that we attain a firm reliance and security with regard to a particular event. Now where is that process of reasoning which, from one instance, draws a conclusion, so different from that which it infers from a hundred instances that are nowise different from that single one? This question I propose as much as for the sake of information, as with an intention of raising difficulties. I cannot find, I cannot imagine any such reasoning. But I keep my mind still open to instruction, if any one will vouchsafe to bestow it on me.

Should it be said that, from a number of experiments, we *infer* a connection between sensible qualities and secret powers; this, I must confess, seems the same difficulty, couched in different terms. The question still recurs, on what process of argument this *inference* is founded? Where is the medium, the interposing ideas, which join propositions so very wide of each other? It is confessed that the colour, consistence, and other sensible qualities of bread appear not, of themselves, to have any connexion with the secret powers of nourishment and support. For otherwise we could infer these secret powers from the first appearance of these sensible qualities, without the aid of experience; contrary to the sentiment of philosophers, and contrary to plain matter of fact. Here, then, is our natural state of ignorance with regard to the powers and influence of all objects. How is this remedied by experience? It only shows us a number of uniform effects, resulting from certain objects, and teaches us that those particular objects, at that particular time, were endowed with such powers and forces. When a new object, endowed with similar sensible qualities, is produced, we expect similar powers and forces, and look for a like effect. From a body of like colour and consistence with bread we expect like nourishment and support. But this surely is a step or progress of the mind, which wants to be explained. When a man says, *I have found, in all past instances, such sensible qualities conjoined with such secret powers*: And when he says, *similar sensible qualities will always be conjoined with such secret powers*; he is not guilty of a tautology, nor are these propositions in any respect the same. You say that the one proposition is an inference from the other. But you must confess that the inference is not intuitive: neither is it demonstrative: Of what nature is it then? To say that it is experimental, is begging the question. For all inferences from experience suppose, as their foundation, that the future will resemble the past, and that similar powers will be conjoined with similar sensible qualities.

If there be any suspicion that the course of nature may change, and that the past may be no rule for the future, all experience becomes useless, and can give rise to no inference or conclusion. It is impossible, therefore, that any arguments from experience can prove this resemblance of the past to the future; since all these arguments are founded on the supposition of that resemblance. Let the course of things be allowed hitherto ever so regular; that alone, without some new argument or inference, proves not that, for the future, it will continue so. In vain do you pretend to have learned the nature of bodies from your past experience. Their secret nature, and consequently all their effects and influence, may change, without any change in their sensible qualities. This happens sometimes, and with regard to some objects: Why may it not happen always, and with regard to all objects? What logic, what process or argument secures you against this supposition? My practice, you say, refutes my doubts. But you mistake the purport of my question. As an agent, I am quite satisfied in the point; but as a philosopher, who has some share of curiosity I will not say scepticism, I want to learn the foundation of this inference. I must confess that a man is guilty of unpardonable arrogance who concludes, because an argument has escaped his own investigation, that therefore it does not really exist. I must also confess that, though all the learned, for several ages, should have employed themselves in fruitless search upon any subject, it may still, perhaps, be rash to conclude positively that the subject must, therefore, pass all human comprehension. Even though we examine all the sources of knowledge, and conclude them unfit for such a subject, there may still remain suspicion, that the enumeration is not complete, or the examination not accurate. But with regard to the present subject, there are some considerations which seem to remove all this accusation of arrogance or suspicion of mistake.

It is certain that the most ignorant and stupid peasants — nay infants, nay even brute beasts — improve by experience, and learn the qualities of natural objects, by observing the effects which result from them. When a child has felt the sensation of pain from touching the flame of a candle, he will be careful not to put his hand near the candle; but will expect a similar effect from a cause which is similar in its sensible qualities and appearances. If you assert, therefore, that the understanding of the child

> is led into this conclusion by any process of argument or ratiocination, I may justly require you to produce that argument; nor have you pretence to refuse so equitable a demand. You cannot say that the argument is abstruse, and may possibly escape your enquiry; since you confess that it is obvious to the capacity of a mere infant. If you hesitate, therefore, a moment, or if, after reflection, you produce an intricate or profound argument, you, in a manner, give up the question, and confess that it is not reasoning which engages us to suppose the past resembling the future, and to expect similar effects from causes which are, to appearance, similar. This is the proposition which I intended to enforce in the present section. If I be right, I pretend not to have made any mighty discovery. And if I be wrong, I must acknowledge myself to be indeed a very backward scholar; since I cannot now discover an argument which, it seems, was perfectly familiar to me long before I was out of my cradle.

The earth revolves about the sun in an elliptical orbit in our solar system, but it may be that planets in other solar systems have different orbits. Newton thought that the sun generated heat by combustion because in his experience, flames, heat and light are only generated by combustion. However, Newton did not have a problem with the fallibility of inductive reasoning, whereas, in the above, Hume argues against the possibility of inductive reasoning — he argues that inductive knowledge was not even probable knowledge.

Arguments against induction are sometimes expressed in terms of the past and the future because experience is always in the past. Hume would tell us that we cannot simply assume that on the basis of no evidence that the future will resemble the past because that is the point in question. Bertrand Russell (p536) explains Hume's thinking:

> Hume is not content with reducing the evidence of a causal connection to experience of frequent conjunction; he proceeds to argue that such experience does not justify the expectation of similar conjunctions in the future. For example: when (to repeat a former illustration) I see an apple, past experience makes me expect that it will taste like an apple, and not like roast beef; but there is no rational justification for this expectation. If there were such a justification it would have to proceed from the principle "that those instances, of which we have had no experience, resemble those of which we have had experience". This principle is not logically necessary, since we can at least conceive a change in the course of nature. It should therefore be a principle of probability. But all probable arguments assume this principle, and therefore it cannot itself be proved to be probable by any such argument. "The supposition, *that the future resembles the past*, is not founded on arguments of any kind, but is derived from habit." The conclusion is one of complete scepticism:

In other words, we cannot use a sample of situation A (the past) to tell us what might be happening in situation B (the future), except if we assume that situation B is the same as or is representative of situation A. Probability requires that the sample reflects the population but if that were the case then we would not need inductive thinking.

Suppose that because of our lack of imagination we said that certain fruit trees fruit at approximately the same time each year — say in December. This induction would be successful for our hemisphere of earth. If we shifted to the other hemisphere we would discover the error in our theory and then we would need to come up with an explanation for why one hemisphere was six months different from the other hemisphere. However, even before we made this discovery by experiment, if someone suggested that season (say ambient temperature and solar radiation) was the true causal variable and that time was only the sensible quality, then we would have two laws/explanations that fit the same pattern. Both laws accurately describe the pattern for the known portion of the pattern (our experience) but they do not always give the same predictions. Both laws accurately describe the pattern for the known portion of the pattern because there is a happen-to-be association between time and season when we remain in the same hemisphere.

In this example of fruiting trees, we see that this induction seems valid and then it fails because we notice that there is a second variable that can be associated/correlated with our pattern and that the second variable is not going to correlate with the predicted pattern if the next instance is for a different hemisphere of the earth. This seems to suggest that induction only works if we are naive. For example, (in an abstract non-empirical sense — that is ignoring that we think we understand which variables are relevant to fruit trees) how do we know that there are not other variables that happen to be correlated that we do not notice — variables that may not be correlated with the pattern in the next instance that someone else might notice but we don't? This is an important part of Hume's problem. Variables are either Hume's secret/true causes (if they are the correct variable) or the sensible qualities (if they are just variables that we notice that happen to correlate with the pattern). Hume said that even though we might correlate a pattern with some sensible quality, the correlation may only be a consequence of the fact that the secret/true variable happens to correlate with the sensible quality for our experience. We know that December is only a sensible quality and that it would correlate with the pattern providing we remained in the same hemisphere.

Or suppose that after some experience we associate smoking with lung cancer. We have sufficient data to be sure this correlation is not a coincidence and in order to *explain* this correlation we might say that smoking causes lung cancer. But the real cause for the law may be that people who smoke are strongly represented by people who have less regard for their health (which is another association) and therefore they may eat more unhealthy food. If this was true then smoking would be what Hume calls the sensible quality. If something changes in the lives of those who smoke (such as their diet) then the association could fail.

Hume attempted to defeat any supposed logical basis for induction in science, without rejecting the idea of a principle of cause and effect and without rejecting the idea that there were permanent associations. Although Hume says things like, "If we would satisfy ourselves, therefore, concerning the nature of that evidence, which assures us of matters of fact, we must enquire how we arrive at the knowledge of cause and effect" (see section IV, part 1, paragraph 5), he seems to be questioning how we should verify the particular cause and effect relationship rather than questioning that there is such a thing as 'cause and effect'. A principle of cause and effect or rather a principle of repeatability merely says that the same effect always follows the cause. Even allowing that there are permanent laws in operation, Hume challenged the idea that a particular example of a supposed law was a true law. That is, Hume rejected that we can use induction with a principle of repeatability in a rational manner in order to discover a specific cause and effect law because he considered that we can never determine the secret causes. The important point here is that even though Russell's and Popper's problem of induction is contained in Hume's argument, the emphasis of Hume's argument was quite different from the modern problem of induction.

Hume argued in an abstract sense that a pattern did not prove that we would always get a pattern. We might say that if we push a button on a machine and get a continuous series of green lights then this does not provide any evidence that we would continue to get green lights. Second, Hume argued that we could use the idea of happen-to-be associations of a secret power/cause with a sensible quality to provide a practical explanation for apparent laws. This introduces the idea of pseudo-laws and prevents us from using a faith in laws to escape his scepticism. Understanding the basis of Hume's reasoning helps us to understand that Hume argued against induction without the use of ultra-sceptical arguments. For example, in section IV, part 2, paragraph 8 Hume said: "This happens sometimes, and with regard to some objects: Why may it not happen always, and with regard to all objects?"

In order to get from the practical problem of associations to the ultra-sceptical problem of colliding billiard balls, Hume maintained that there is no way we can determine if a pattern that seems to be a true law is a true law or a pseudo-law because the only tool we have at our disposal is experience. The conclusion is that we only trust things like colliding billiard balls because we are very familiar with colliding billiard balls. Effectively we know what is going to happen so we think we understand. But, if we truly understood colliding billiard balls then we would not need experience to tell us what is going to happen. According to Hume's thinking, if we truly understood gravity then we should not be able to imagine an object falling upwards just as we cannot imagine how the laws of logic might be violated.

As previously suggested we need to divide Hume's argument into two slightly different arguments. There is the modern day Popper/Russell/Hume problem of induction concerned with establishing why we think that a pattern that has been true should remain true. Why should we expect all swans to be white? Despite the fact that we can verify our description of the known pattern, we can also imagine many other descriptions (or hypotheses) that comply with the known pattern but give a different description for the unknown part of the pattern. Or, from another direction, we could say that even if there were only one pattern that describes the past, who is to say that the future pattern will be the same? In these arguments, we are focused on the pattern and have no interest in variables and true causes. Such arguments are really about Bacon's induction by enumeration. Patterns such as 'the sun rising each day' and that 'objects are conserved' naturally fit this category because it is difficult to imagine which variables might affect the pattern.

However, Hume was more interested in a slightly different problem. Using Hume's terminology we would say that he was questioning the idea that we could discover the laws even if we know that there are true laws. This is because we could never know the secret causes (the true and permanent associations) and therefore the secret causes may change independently of the sensible qualities we wishfully associate with the causes. Sensible qualities are qualities that strike our attention and any other qualities we may be able to observe; Hume's point is that for all we know, all such qualities are arbitrary qualities. This is the problem of associations or the problem of empirical induction with the main issue being the possibility of pseudo-variables and temporary associations. The following is a summary of Hume's argument against empirical induction:

1. We can never truly understand empirical laws.
2. Our only way of learning the behaviour of empirical things is by observation.
3. We note repetitive associations between factors such as smoking and cancer.

4. We explain our associations using cause and effect — we say that smoking causes cancer.

5. We think that we have an understanding because we can make predictions and because we think we have understanding, we think that we have identified a law. Alternatively, we could say that we think that we have identified a law because a law explains the consistency.

6. For all we know we may have only discovered a temporary association. A temporary association will always explain a finite portion (which is all we ever observe) of a permanent association.

7. It does happen sometimes that our patterns fail. It does happen that some of our patterns turn out to be temporary associations so it is possible that this is the case for any particular association.

8. We cannot determine the true causes so we can never be certain that we have a law.

9. Because we can only learn by experience, there is no distinction between complex association problems and clear and simple problems when we think we understand *therefore empirical induction does not work.*

10. That empirical induction does not work means we must fall back on induction by enumeration, but induction by enumeration cannot work either because induction by enumeration is empirical induction without attempting to identify the variables. Instead of saying ‘induction does not work’ philosophers prefer to say ‘induction is not justifiable.’

1.2 Mill

The fundamental reason why Hume argued that we cannot determine probable empirical causes was that we can never determine the true variables. John Stuart Mill (1806-1873), a contemporary of Charles Darwin, pointed out in his *System of Logic* that we can question nature by doing things like changing one variable and keeping everything else constant to prove that the variable is a true fundamental causal variable — it is not just an association. All of Mill's arguments cited in this section come from Mills *System of Logic.*

Experiment is a special type of experience that we all use that is not induction by enumeration and this allows us to avoid Hume's temporary patterns. The principle of cause and effect (or the principle of repeatability) says that the associations between causes and effects do not vary. Alternatively, we say that when we repeat an experiment we will get the same results. This means that once we have discovered the variables then the law will be permanent providing we can trust the principle of cause and effect.

When we can control the variables in order to determine the true variables, then we do not need induction. Mill's argument does not actually address the problem of induction as expressed by Popper and Russell, but it does address the main part of Hume's argument. In Book 3, chapter 7, sections 2 & 3, Mill explains how we can solve the problem of empirical induction providing we accept the principle of repeatability:

> 2. The different antecedents and consequents being, then, supposed to be, so far as the case requires, ascertained and discriminated from one another, we are to inquire

which is connected with which. In every instance which comes under our observation, there are many antecedents and many consequents. If those antecedents could not be severed from one another except in thought, or if those consequents never were found apart, it would be impossible for us to distinguish (*à posteriori* at least) the real laws, or to assign to any cause its effect, or to any effect its cause. To do so, we must be able to meet with some of the antecedents apart from the rest, and observe what follows from them; or some of the consequents, and observe by what they are preceded. We must, in short, follow the Baconian rule of *varying the circumstances*. This is, indeed, only the first rule of physical inquiry, and not, as some have thought, the sole rule; but it is the foundation of all the rest.

For the purpose of varying the circumstances, we may have recourse (according to a distinction commonly made) either to observation or to experiment; we may either *find* an instance in nature suited to our purposes, or, by an artificial arrangement of circumstances, *make* one. The value of the instance depends on what it is in itself, not on the mode in which it is obtained: its employment for the purposes of induction depends on the same principles in the one case and in the other, as the uses of money are the same whether it is inherited or acquired. There is, in short, no difference in kind, no real logical distinction, between the two processes of investigation. There are, however, practical distinctions to which it is of considerable importance to advert.

3. The first and most obvious distinction between Observation and Experiment is, that the latter is an immense extension of the former. It not only enables us to produce a much greater number of variations in the circumstances than nature spontaneously offers, but,

also, in thousands of cases, to produce the precise *sort* of variation which we are in want of for discovering the law of the phenomenon, — a service which nature, being constructed on a quite different scheme from that of facilitating our studies, is seldom so friendly as to bestow upon us. For example, in order to ascertain what principle in the atmosphere enables it to sustain life, the variation we require is that a living animal should be immersed in each component element of the atmosphere separately. But nature does not supply either oxygen or azote in a separate state. We are indebted to artificial experiment for our knowledge that it is the former, and not the latter, which supports respiration; and for our knowledge of the very existence of the two ingredients. Thus far the advantage of experimentation over simple observation is universally recognised: all are aware that it enables us to obtain innumerable combinations of circumstances which are not to be found in nature, and so add to nature's experiments a multitude of experiments of our own. But there is another superiority (or, as Bacon would have expressed it, another prerogative) of instances artificially obtained over spontaneous instances, — of our own experiments over even the same experiments when made by nature, — which is not of less importance, and which is far from being felt and acknowledged in the same degree.

When we can produce a phenomenon artificially, we can take it, as it were, home with us, and observe it in the midst of circumstances with which in all other respects we are accurately acquainted. If we desire to know what are the effects of the cause A, and are able to produce A by means at our disposal, we can generally determine at our own discretion, so far as is compatible with the nature of the phenomenon A, the whole of the circumstances which shall be present along with it: and thus, knowing exactly the simultaneous state of everything else which is within

> the reach of A's influence, we have only to observe what alteration is made in that state by the presence of A.

According to Mill's thinking, if we have a set of potential causes of a phenomenon (say a car not starting) but we find that when we remove only one of the potential causes (say by replacing the spark plugs and keeping everything else constant) and the car starts, then we have found a cause (or part of the cause) of the phenomenon.

The catch to Mill's solution is that to assume that there is such a thing as a principle of cause and effect then we must first accept inductive reasoning (induction by enumeration or what Bacon called childish induction) because this is the only way we get to learn the principle of cause and effect. Although Hume seemed happy to allow that there was such a thing as a principle of cause and effect, perhaps that merely means that Hume wanted to avoid super-sceptical arguments.

It could be argued that Hume only accepted a principle of cause and effect for the purposes of his argument, not caring that the reliability of the principle of cause and effect is based on induction. That is, if Hume could show that empirical induction cannot be validated (because we do not know that we have the right variables) even if there is a principle of cause and effect, then it does not really matter that the principle of cause and effect is itself questionable. In such a case, Hume would have two reasons for why he should not expect an empirical pattern to continue: he could say that 'we do not know that we have determined the correct *relationship* of cause and effect' or he could say that 'we do not know that there is such a thing as cause and effect and therefore for all we know there is no correct relationship'.

Hume's argument rested on the idea that we can never determine the variables but that we only ever observe a consistent pattern. As already discussed in section IV, part

2, Hume added the understanding that the pattern sometimes fails. Hume said, "This happens sometimes, and with regard to some objects: Why may it not happen always, and with regard to all objects?" Hume said that we could easily explain patterns through the use of temporary associations and their consequent pseudo-laws and that we know from experience that such temporary associations do occur and that our laws sometimes fail. We know, for example, that the theory that fruit trees fruit in December can fail regardless of how many times it has been confirmed by experience. Although, there are still some things to be resolved, we can say that *for whatever reason*, an argument against the principle of cause and effect would require a different level of scepticism. If we want to be sceptical, we could say that we have no reason to think that when we put a book on the desk that some part of it might change or disappear. Or, we could say that (other than from experience) we have no reason to think that when we rearrange 20 bricks, that after rearranging the bricks, the number of bricks will not correspond to the number 20 and therefore, there is no reason to think that number theory works in practice.

Although Hume tried to avoid super-sceptical arguments, he nevertheless came to super-sceptical conclusions when he extended the associations problem to simple situations such as colliding billiard balls. But, when Hume extended his practical arguments to simple situations, he failed to realise that we can eliminate pseudo-variables. While it is true that there may be other variables associated with two colliding billiard balls that we cannot even imagine, we can again say that such an argument requires a different level of scepticism than the argument 'that for all we know there might be some other causal variable associated with smoking'. We can therefore say that Mill solved Hume's main argument. Mill provided the solution to Hume's problem of *empirical* induction.

Mill has shown us that one of Hume's errors was to fail to see how we can eliminate pseudo-variables. In fact Hume's assertion of counter-instances did not add anything to the argument, because we know that we are going to get situations of temporary associations when the situations are complicated (such as the case of smoking causing lung cancer). We know that there is a problem of how to use empirical induction that concerns 'secret causes and sensible qualities'.

Mill removed the strength of Hume's argument and left only super-sceptical arguments. Mill showed us that we can eliminate Hume's temporary associations and secret powers. We can do that if we believe in repeatability. While it is true that induction by enumeration is only empirical induction when we do not bother to determine the variables for whatever reason, only super-sceptical thinking can doubt things like the principle of repeatability and the law of conservation of objects.

Contrary to Hume's understanding of necessary truth, Mill said that habitual association gives experimental truth the appearance of a necessary truth. In Book 2, chapter 5, section 4, Mill says of Euclidian geometry and induction by enumeration: " § 4. [Those first principles of geometry that are axioms are experimental truths] It remains to inquire, what is the ground of our belief in axioms—what is the evidence on which they rest? I answer, they are experimental truths; generalizations from observation. The proposition, Two straight lines cannot inclose a space—or in other words, Two straight lines which have once met, do not meet again, but continue to diverge—is an induction from the evidence of our senses." In Book 2, chapter 6, section 2, Mill says something similar of number theory: "The Science of Number is thus no exception to the conclusion we previously arrived at, that the processes even of deductive sciences are altogether inductive, and that their first principles are generalizations from experience. It remains to be examined whether this science resembles geometry in the further circumstance, that some of its inductions are not exactly true; and that the peculiar certainty ascribed to it, on account of which its propositions are called Necessary Truths, is fictitious and hypothetical, being true in no other sense than that those propositions legitimately follow from the hypothesis of the truth of premises which are avowedly mere approximations to truth."

In Book 2, chapter 5, section 6 of his *System of Logic*, Mill argues against William Whewell's (1794-1866) 'conceivable in the mind criterion', which was the criterion that Hume used to distinguish empirical truth from analytic truth:

> Although Dr. Whewell [a contemporary of Mill] has naturally and properly employed a variety of phrases to bring his meaning more forcibly home, he would, I presume, allow that they are all equivalent; and that what he means by a necessary truth, would be sufficiently defined, a proposition the negation of which is not only false but inconceivable. I am unable to find in any of his expressions, turn them what way you will, a meaning beyond this, and I do not believe he would contend that they mean anything more.
>
> This, therefore, is the principle asserted: that propositions, the negation of which is inconceivable, or in other words, which we cannot figure to ourselves as being false, must rest on evidence of a higher and more cogent description than any which experience can afford. Now I cannot but wonder that so much stress should be laid on the circumstance of inconceivableness, when there is such ample experience to show, that our capacity or incapacity of conceiving a thing has very little to do with the possibility of the thing in itself; but is in truth very much an affair of accident, and depends on the past

history and habits of our own minds. There is no more generally acknowledged fact in human nature, than the extreme difficulty at first felt in conceiving anything as possible, which is in contradiction to long established and familiar experience; or even to old familiar habits of thought. And this difficulty is a necessary result of the fundamental laws of the human mind. When we have often seen and thought of two things together, and have never in any one instance either seen or thought of them separately, there is by the primary law of association an increasing difficulty, which may in the end become insuperable, of conceiving the two things apart. This is most of all conspicuous in uneducated persons, who are in general utterly unable to separate any two ideas which have once become firmly associated in their minds; and if persons of cultivated intellect have any advantage on the point, it is only because, having seen and heard and read more, and being more accustomed to exercise their imagination, they have experienced their sensations and thoughts in more varied combinations, and have been prevented from forming many of these inseparable associations. But this advantage has necessarily its limits. The most practised intellect is not exempt from the universal laws of our conceptive faculty. If daily habit presents to any one for a long period two facts in combination, and if he is not led during that period either by accident or by his voluntary mental operations to think of them apart, he will probably in time become incapable of doing so even by the strongest effort; and the supposition that the two facts can be separated in nature, will at last present itself to his mind with all the characters of an inconceivable phenomenon.* There are remarkable instances of this in the history of science: instances in which the most instructed men rejected as impossible, because inconceivable, things which their posterity, by

> earlier practice and longer perseverance in the attempt, found it quite easy to conceive, and which everybody now knows to be true. There was a time when men of the most cultivated intellects, and the most emancipated from the dominion of early prejudice, could not credit the existence of antipodes; were unable to conceive, in opposition to old association, the force of gravity acting upwards instead of downwards. The Cartesians [followers of Descartes] long rejected the Newtonian doctrine of the gravitation of all bodies towards one another, on the faith of a general proposition, the reverse of which seemed to them to be inconceivable—the proposition that a body cannot act where it is not. All the cumbrous machinery of imaginary vortices, assumed without the smallest particle of evidence, appeared to these philosophers a more rational mode of explaining the heavenly motions, than one which involved what seemed to them so great an absurdity. And they no doubt found it as impossible to conceive that a body should act upon the earth from the distance of the sun or moon, as we find it to conceive an end to space or time, or two straight lines inclosing a space. Newton himself had not been able to realize the conception, or we should not have had his hypothesis of a subtle ether, the occult cause of gravitation; and his writings prove, that though he deemed the particular nature of the intermediate agency a matter of conjecture, the necessity of some such agency appeared to him indubitable.

It could be argued that Mill agrees with Hume that there is no logical basis for induction by enumeration because Mill argues that even our trust in what seems to be necessary truth comes from experience. It could be argued that Mill agrees with Russel that induction is an independent principle. Or, it could be argued that Mill's 'primary law of association' agrees with Aristotle's psychological explanation for induction namely that: 'out of sense-perception comes to be what we call memory, and out of frequently repeated memories of the same thing develops experience; for a number of memories constitute a single experience... for the method by which even sense-perception implants the universal is inductive'.

Even though we know God can do what he likes in his universe, there seems to be no doubt that Mill is correct to suggest that we get psychological 'certainty' from experience. We trust that '10 bricks + 10 bricks = 20 bricks' and *whether or not there is or is not a logical basis for such a trust* there must also be a psychological one that could be explained in terms of Mill's primary law of association. It must be that such laws become part of our mental makeup in the sense that we will automatically assume the laws are true without ever recognising the assumption because that is what we all do.

Mill solved Hume's problem of empirical induction. Modern clinical trials of drugs is an example of how science recognises and addresses the problem of empirical induction. Clinical trials of drugs randomly assign drugs to eliminate (or rather probably eliminate) Hume's temporary associations. The solution to Hume's problem of induction can be found in science that came before Hume. Newton used Mill's thinking when he questioned the variables. According to Mill, the methods of Francis Bacon contained the solution to Hume's main problem. And according to Francis Bacon, even the methods of Plato contained the solution to Hume's problem of induction. However, to truly solve Hume's problem, we must first take on Hume's focus and then solve Hume's problem, which is what Mill did.

In summary, we can say that Hume's thoughts were very much focused on the idea that we can only learn about the real world by experience and induction, which *seems* to be an obvious truth. To Hume, this was the beginning and the end of the problem which seems to be a very fair argument from an abstract perspective. However, Mill said we could learn by questioning nature and elimination of pseudo-variables, which must also be true. Both ideas are correct, even though they seem to be contradictory. We can only learn by induction but when we learn the principle of repeatability by induction (which we all think that we have done), we can question nature using Mill's methods. Mill rectified one of Hume's errors, solving an important part of Hume's problem of induction, which is a different problem of induction to that identified by Russell and Popper (and Hume as the background to his overall argument) .

1.3 Popper

Like Russell, Popper ignores Mill's solution. Popper is focused on induction by enumeration. Apparently neither Popper nor Russell accept what Mill has to say about induction by enumeration.

Induction by enumeration is concerned with why we can trust the conservation of objects and why we should trust that the sun will rise tomorrow and why we should trust that things will continue to fall down. Induction by enumeration is more about why a pattern should remain constant rather than about true causes, sensible qualities and secret powers.

Popper objects to arguments that say that we can trust induction because 'it works'. Experience is always past tense. A reference to experience as proof that induction works is to say that we know the pattern will be constant because it has been constant. On pages 3-7 of his *Logic of Scientific Discovery* Popper argues that his testability criterion avoids the problem of induction:

> According to a widely accepted view — to be opposed in this book — the empirical sciences can be characterised by the fact that they use '*inductive methods*', as they are called. According to this view, the logic of scientific discovery would be identical with inductive logic, i.e. with the logical analysis of these inductive methods.
>
> It is usual to call an inference 'inductive' if it passes from *singular statements* (sometimes also called 'particular' statements), such as accounts of the results of observations or experiments, to *universal statements*, such as hypotheses or theories.

Now it is far from obvious, from a logical point of view, that we are justified in inferring universal statements from singular ones, no matter how numerous; for any conclusion drawn in this way may always turn out to be false: no matter how many instances of white swans we may have observed, this does not justify the conclusion that all swans are white.

The question whether inductive inferences are justified, or under what conditions, is known as *the problem of induction.* The problem of induction may also be formulated as the question of the validity or the truth of universal statements which are based on experience, such as the hypotheses and theoretical systems of the empirical sciences. For many people believe that the truth of these universal statements is '*known by experience*'; yet it is clear that an account of experience — of an observation or the result of an experiment — can in the first place be only a singular statement and not a universal one. Accordingly, people who say of a universal statement that we know its truth from experience usually mean that the truth of this universal statement can somehow be reduced to the truth of singular ones, and that these singular ones are known by experience to be true; which amounts to saying that the universal statement is based on inductive inference. Thus to ask whether there are natural laws known to be true appears to be only another way of asking whether inductive inferences are logically justified.

Yet if we want to find a way of justifying inductive inferences, we must first of all establish a *principle of induction.* A principle of induction would be a statement with the help of which we could put inductive inferences into a logically acceptable form. In the eyes of the upholders of inductive logic, a principle of induction is of supreme importance for scientific method:'...this principle', says Reichenbach, 'determines the truth of scientific theories. To eliminate it from science would mean nothing less than to deprive science of the power to decide the truth or falsity of its theories. Without it, clearly, science would no longer have the right to distinguish its theories from the fanciful and arbitrary creations of the poet's mind.'

Now this principle of induction cannot be a purely logical truth like a tautology or an analytic statement. Indeed, if there were such a thing as a purely logical principle of induction, there would be no problem of induction; for in this case, all inductive inferences would have to be regarded as purely logical or tautological transformations, just like inferences in deductive logic. Thus the principle of induction must be a synthetic statement; that is, a statement whose negation is not self-contradictory but logically possible. So the question arises why such a principle should be accepted at all, and how we can justify its acceptance on rational grounds. Some who believe in inductive logic are anxious to point out, with Reichenbach, that 'the principle of induction is unreservedly accepted by the whole of science as that no man can seriously doubt this principle in everyday life either'. Yet even supposing this were the case — for after all, 'the whole of science' might err — I should still contend that a principle of induction is superfluous, and that it must lead to logical inconsistencies.

That inconsistencies may easily arise in connection with the principle of induction should have been clear from the work of Hume; also, that they can be avoided, if at all, only with difficulty. For the principle of induction must be a universal statement in its turn. Thus if we try to regard its truth as known from experience, then the very same problems which occasioned its introduction will arise all over again. To justify it, we should have to employ inductive inferences; and to justify these we should have to assume an inductive principle of a higher order; and so on. Thus the attempt to base the principle of induction on experience breaks down, since it must lead to an infinite regress.

Kant tried to force his way out of this difficulty by taking the principle of induction (which he formulated as the 'principle of universal causation') to be '*a priori* valid'. But I do not think that his ingenious attempt to provide an *a priori* justification for synthetic statements was successful. My own view is that the various difficulties of inductive logic here sketched are insurmountable. So also, I fear, are those inherent in the doctrine, so widely current today, that inductive inference, although not 'strictly valid', *can attain some degree of 'reliability' or of 'probability'*. According to this doctrine, inductive inferences are 'probable inferences'. "We have described", says Reichenbach, "the principle of induction as the means whereby science decides upon truth. To be more exact, we should say that it serves to decide upon probability. For it is not given to science to reach either truth or falsity...but scientific statements can only attain continuous degrees of probability.

At this stage I can disregard the fact that the believers in inductive logic entertain an idea of probability that I shall later reject as highly unsuitable for their own purposes (see section 80, below). I can do so because the difficulties mentioned are not even touched by an appeal to probability. For if a certain degree of probability is to be assigned to statements based on inductive inference, then this will have to be justified by invoking a new principal of induction, appropriately modified. And this new principle in its turn will have to be justified, and

> so on. Nothing is gained, moreover, if the principle of induction, in its turn, is taken not as 'true' but only as 'probable'. In short, like every other form of inductive logic, the logic of probable inference, or 'probability logic', leads either to an infinite regress, or to the doctrine of *apriorism*. The theory to be developed in the following pages stands directly opposed to all attempts to operate with the ideas of inductive logic. It might be described as the theory of *the deductive method of testing* , or as the view that a hypothesis can only be empirically tested — and only *after* it has been advanced.

Because Popper is focused on universal laws, it is natural that he is interested in statements like 'numerous confirming instances of white swans do not tell us that all swans are white'. However, as can be seen in the notes Popper Rejects Any Possibility of Inductive Inferences. On pages 641 and 646-647 of his History of Western Philosophy, Bertrand Russell argues:

> These two parts of the doctrine may be stated as follows: (1) in causation there is no indefinable relation except conjunction or succession; (2) induction by simple enumeration is not a valid form of argument. Empiricists in general have accepted the first of these theses and rejected the second. When I say they have rejected the second, I mean that they have believed that, given a sufficiently vast accumulation of instances of a conjunction, the likelihood of the conjunction being found in the next instance will exceed half; or, if they have not held exactly this, they have maintained some doctrine having similar consequences.
>
> I do not wish, at the moment, to discuss induction, which is a large and difficult subject; for the moment, I am content to observe that, if the first half of Hume's doctrine is admitted, the rejection of induction makes all expectation

as to the future irrational, even the expectation that we shall continue to feel expectations. I do not merely mean that our expectations may be mistaken; that, in any case, must be admitted. I mean that, taking even our firmest expectations, such as that the sun will rise tomorrow, there is not a shadow of reason, for supposing them more likely to be verified than not...

It is therefore important to discover whether there is any answer to Hume within the framework of a philosophy that is wholly or mainly empirical. If not, there is no intellectual difference between sanity and insanity. The lunatic who believes that he is a poached egg is to be condemned solely on the ground that he is a minority, or rather — since we must not assume democracy — on the ground that the government does not agree with him. This is a desperate point of view, and it must be hoped that there is some way of escaping from it.

Hume's scepticism rests entirely upon his rejection of the principle of induction. The principle of induction, as applied to causation says that, if A has been found very often accompanied or followed by B, and no instance is known of A not being accompanied or followed by B, then it is probable that on the next occasion on which A is observed it will be accompanied or followed by B. If the principle be adequate, a sufficient number of instances must make the probability not far short of certainty. If this principle, or any other from which it can be deduced, is true, then the causal inferences which Hume rejects are valid, not indeed as giving certainty, but as giving a sufficient probability for practical purposes. If this principle is not true, every attempt to arrive at general scientific laws from particular observations is fallacious, and Hume's scepticism is inescapable for an empiricist. The principle itself cannot, of course, without circularity,

> be inferred from observed uniformities, since it is required to justify any such inference. It must therefore be, or be deduced from, an independent principle not based upon experience. To this extent, Hume has proved empiricism is not a sufficient basis for science. But if this one principle is admitted, everything else can proceed in accordance with the theory that all our knowledge is based on experience. These, however are questions not directly raised by Hume's arguments. What these arguments prove — and I do not think the proof can be controverted — is, that induction is an independent logical principle, incapable of being inferred either from experience or from other logical principles, and that without this principle science is not possible.

If, whenever we push a button on some sort of sequence machine, we note that a green light comes on — how do we justify the idea that the green light will always come on? We cannot say that 'because it always came on before it will come on again' because this is the argument that we are trying to justify. It may be that we tend to believe induction will work because it has worked, because from our experience, we note that things tend to be uniform or because laws seem to regulate behaviour. But, even though induction has worked (past tense), there is a problem in saying that it will work (future tense) because this conclusion itself relies on the idea that induction works. Experience is always past tense and we want it to tell us something about the future without simply assuming induction works. Reverting to more general experience (rather than the specific apparent law we are interested in) is similar to the case of having lots of different predictable sequence machines. We cannot use the idea that they all seem to be predictable to justify that induction works for a particular machine because again, this is the type of argument we are trying to justify.

In summary, we see that Hume claimed that induction could not be justified from logic, but he believed that we still successfully use it as some sort of mental process. Mill solved Hume's problem of empirical induction and left us with the problem of induction by enumeration (or simply the problem of induction). Popper claimed that induction is invalid and that, in fact, we never use it. He believed that the process we use is actually 'a trial-and-error process that looks very much like induction'. Russell claimed that induction must be some independent principle which is not justifiable from either logic or experience. Mill claimed that we use induction to justify other principles. And even though there seems to be no doubt that there is such a thing as a primary law of association that teaches us to trust familiar patterns, there is still the question of the logical justification for 'induction by enumeration'.

Chapter 2 JUSTIFYING INDUCTION

Hume questioned the reliability of cause-and-effect patterns because he claimed such patterns could only be known by induction, but Mill effectively said we use experiment coupled with a reliance on repeatability (empirical induction), rather than induction, to determine such patterns. Mill said we can change some supposed variables while keeping other things constant to see if the supposed variable is a true variable. According to Mill's thinking, in the case of the fruit trees being alleged to fruit in December, we should experiment to see if December or summer is the true causal variable. Mill's strategy is based on the idea of elimination of pseudo-variables so that we can be sure that we have found the true variables — assuming we can rely on the principle of repeatability and assuming that we can guess the potential variables. Mill not only argued that it is a part of our mental makeup to trust consistent patterns (induction by enumeration) but he also argued that Hume's 'inconceivable in the mind' state comes from frequent observation of consistent patterns.

Of course it is true that we make a countless number of background assumptions during the normal course of life. We assume gravity is going to keep us on the earth and we assume that the earth is not going to collapse beneath our feet and that our heart is going to keep beating, etc. Such assumptions are more like foregone conclusions that are hidden from view and in the interest of speed and efficiency they are never deliberated. We need something like Mill's psychological explanation to explain our trust in such hidden assumptions.

However, there is still the question as to whether it is possible to justify induction (or rather 'induction by enumeration') with logic. If we can justify our trust in induction with logic then we justify our trust in the principle of repeatability, conservation of objects and many other such patterns with logic. And on the other hand, because we know that God can do what he likes in the empirical world, then for sure, we know that it is impossible to justify a universal law. Given that it is impossible to justify a universal law it must help if we frame the problem of induction so that it excludes universal laws. The following arguments are based on two different kinds of inductive inferences, one that is about the justification of universal laws and one that is about the justification of a near instance:

- *Type 1 Induction*: Suppose we have a known pseudo-law — a temporary association or a pattern which we know is going to sooner or later fail. We will call this a type 1 pattern. A type 1 problem of induction is the justification of a type 1 pattern for the near instances — let's say for the next instance. We must choose near instances since the far/infinite instances are excluded by definition. An inductive conclusion for a near instance can be called a type 1 inference.
- *Type 2 Induction*: A universal law or a pattern that is always true can be called a type 2 pattern. The type 2 problem of induction is concerned with the justification of type 2 patterns. In this case we want to be able to say that a law will always be true. Popper would say 'that all swans have been white does not prove that all swans are white'. Popper also makes the claim that the probability of any universal law is zero. Because we know that God can do what he likes in the empirical world then, it will never be possible to absolutely trust a type 2 empirical pattern.

Although the type 2 problem is the main problem of Popper's philosophy of science and it is the problem that testability supposedly addresses, most of what is important in this chapter concerns justification of type 1 inferences (section 2.1), because type 1 inferences are the only kind of inferences that can be justified. Section 2.2 briefly considers Popper's comparison of the laws (type 2 patterns) of science to unproved laws of mathematics. Section 2.3 briefly considers Popper's discussion of the simplicity in connection with mathematical curves, and other issues that have a bearing on the reliability of empirical induction. Although such issues sometimes find their way into a discussion of the problem of induction, almost all confusion concerning the problem of induction comes from the validity of type 1 induction.

2.1 Type 1 Induction

If we have a long pattern but we know it is going to run out at some finite time (a type 1 pattern), we may want to be able to attach some reliability to the next instance. According to Hume's thinking, even though we have a suspicion that the pattern might go on for a long time, there is no way of justifying a probable *inference to the next instant.* This argument maintains that *induction should not be logically useful* even to say that the next swan should probably be white. The alternative view is that induction is a logical process but it's justification might be elusive. Hume (*Enquiry* Part II) considered the argument that 'the justification of induction might be elusive':

> I must confess that a man is guilty of unpardonable arrogance who concludes, because an argument has escaped his own investigation, that therefore it does not really exist. I must also confess that, though all the learned, for several ages, should have employed themselves in fruitless search upon any subject, it may still, perhaps, be rash to conclude positively that the subject must, therefore, pass all human comprehension. Even though we examine all the sources of knowledge, and conclude them unfit for such a subject, there may still remain suspicion, that the enumeration is not complete, or the examination not accurate. But with regard to the present subject, there are some considerations which seem to remove all this accusation of arrogance or suspicion of mistake.

The considerations that Hume cites for not believing that there could be a logical justification for induction are because children and animals use induction. Hume's conclusion was that we use induction because of a habit of the mind. Instead of saying habit of the mind we could say instinct or sub-conscious thought processes — this seems like a possibility because small children seem to have a sub-conscious understanding of complex laws of grammar and animals can have sophisticated, instinctive knowledge.

Contrary to what Hume says, such considerations do not in fact remove all accusations of mistake. If induction does not have a logical basis or if its basis passes all human comprehension, then animals and children and adults cannot use it from a logical basis, but if we do not use it from a logical basis this does not mean that a logical basis for trusting patterns does not exist or that it passes all human comprehension. Furthermore, we have already seen that Hume's abstract reasoning failed to identify how we might use elimination (as per Mill's methods) to identify pseudo-variables, even though this is a process that Hume possibly used himself in solving problems. There could be a logical justification for induction that he missed as well.

In Appendix *VII of *Logic of Scientific Discovery*, Popper considers induction from the point of view of two probability expressions, one that he labels '(1)' (which he agrees with) and the other he labels '(+)' (which he does not agree with):

- "(1) $P(a) = 0$ "
- "(+) $P(a_n \mid a_1 a_2 ... a_{n-1}) > P(a_n)$ "

In the above, *P(a)* is a reference to the probability of some conjectured universal law being true, $a_1 a_2 ... a_{n-1}$ are some instances of some law that we have experienced and a_n is a future instance we are trying to predict. The first expression conjectures that the probability of a universal law is zero. The second expression (that Popper does not agree with) conjectures that the probability of a future instance given the evidence of a number of past instances, which conformed to the law, is greater than the probability without such evidence.

On p381-383 of Appendix *VII of *The Logic of Scientific Discovery,* Popper comments on the second expression as follows:

> This view asserts that once we have found some k_i to possess the property *A*, the probability increases that another k_j possesses the same property; and even more so if we have found the property in a number of cases. Or in Hume's terminology, (+) asserts '*that those instances*' (for example k_n), '*of which we have had no experience, are likely to resemble those, of which we have had experience*'.
>
> The quotation, except for the words 'are likely to', is taken from Hume's criticism of induction. And Hume's criticism fully applies to (+), or its italicized verbal formulation. For, Hume argues, '*even after the observation of the frequent constant conjunction of objects, we have no reason to draw any inference concerning any object beyond those of which we have had no experience*'. If anybody should suggest that our experience entitles us to draw inferences from observed to unobserved objects, then Hume says, '*I would renew my question, why from this experience we form any conclusion beyond those past instances, of which we have had experience*'. In other words, Hume points out that we get involved in an infinite regress if we appeal to experience in order to justify any conclusion concerning unobserved instances — *even mere probable conclusions* as he adds in his *Abstract.* For there we read: 'It is evident that Adam, with all his science, would never have been able to *demonstrate* that the course of nature must continue uniformly the same...Nay, I will go further and assert that he could not so much as prove by any *probable* arguments that the future must be conformable to the past. All probable arguments are built on the supposition that there is conformity betwixt the future and the past, and

> therefore can never prove it.' Thus (+) is not justifiable by experience; yet in order to be logically valid, it would have to be of the character of a tautology, valid in every logical possible universe. But this is clearly not the case.
>
> Thus (+), if true, would have the logical character of *a synthetic a priori principle* of induction, rather than of an analytic or logical assertion. But it does not quite suffice even as a principle of induction. For (+) may be true, and $p(a) = 0$ may be valid none the less. (An example of a theory which accepts (+) as a priori valid-though, as we have seen, (+) must be synthetic — and which at the same time accepts $p(a) = 0$, is Carnap's.)
>
> An effective probabilistic principle of induction would have to be even stronger than (+). It would have to allow us, at least, to conclude that for some fitting singular evidence b, we may obtain $p(a \mid b) > 1/2$, or in words, that a may be made, by accumulating evidence in its favour, more probable than its negation. But this is only possible if (1) is false, that is to say, if we have $p(a) > 0$.

Where Popper says that 'in order for induction to be logically valid, it would have to be of the character of a tautology, valid in every logical possible universe' he simply means that there is nothing empirical about induction by enumeration. If induction by enumeration makes sense then it must make sense for the same reason that God has no choice about making little green apples little and green.

Mathematics and logic seems like tautological thinking because mathematics and logic seems like the sort of things that will be true in every possible universe. Probability theory also seems tautological.

Probability theory provides a way of adding one or more decimal places of accuracy to words like, may, should, most likely, etc. When we have a type 1 problem, any conclusion about the next instance cannot be certain, so at most, it can only be probable. This leads us to the problem of identifying what is truly probable and what is not. For all we know, we might be on the edge of the discontinuity and the next instance might give us a failure — if not the next instance then perhaps the one after. Because we are not able to jump to the conclusion that we have a type 2 pattern, we must doubt the future pattern. And given that we have said that we are dealing with a type 1 pattern that is going to run out, the one thing we know for sure is that each instance must be taking us one instance closer to the point of discontinuity in the pattern.

In the following discussion, I will be arguing that 'information probability', which is similar to 'classical probability' and 'logical probability', is the correct perspective for almost all probability values. By 'information probability' all I mean is the view that the probability values for a certain event are values that come from or are conditional upon the information we have about the event. Information probability is not the true propensity of an event to occur and it is not the 'limiting frequency probability'. We use a conditional probability expression when we expect that the probability value might be updated by some new information.

The information view says that apart from quantum mechanics exceptions, the real propensity probability values do not exist except as ideals in our mind. Hume apparently agrees with the information understanding of probability theory. In Section VI of *An Enquiry Concerning Human Understanding*, Hume opens his discussion of probability theory with the statement: "Though there be no such thing as Chance in the world; our ignorance of the real cause of any event has the same influence on the understanding, and begets a like species of belief or opinion" — which is to say, that ignorance generates a concept of probability like the ideal propensity concept of probability. Hume can be excused for not knowing about things like the half-lives of radio-active materials — the probability values of which are apparently truly unconditional probability values (or true propensity values) that can never be updated by new information.

Even though a coin toss is one of simplest probability events and it is one of those events where we are not likely to be confused by new information, people who advocate the 'limiting frequency probability' sometimes say that there is a problem with clearly defining what a probability of 0.5 means for a single situation. The issue is that a probability of 0.5 of getting tails *seems* to lack meaning if we want to say what we should expect — because we are either going to get one outcome or the other and not something halfway between.

The 'limiting frequency method' of attaching meaning to the probability of a coin toss is to toss a coin (or imagine tossing the coin) so many times that probability can make a prediction. Instead of saying that the probability of heads in the next toss is 0.5 we can say that the fraction of heads in the next million tosses of the coin will be 0.500. Despite the fact that there is no conceptual difference between tossing a coin 1,000,000 times or in tossing 1,000,000 coins in one go, or in tossing a single coin once, limiting frequency theory says that probability values make more sense in repetitive situations. However, a probability value of 0.5 for a single event says that the probability is not very high and it is not very low, which seems appropriately deficient in meaning because probability values are about information and a probability that is 0.5 means that we do not know very much.

Suppose we have at our disposal information to allow us to continuously improve our probability value associated with getting certain outcomes from a number of situations — we create situations where can know everything but we choose to limit our knowledge. For a particular set of situations, for which we have only progressed to resolving that the event has a probability of 0.5, we can say that we have not resolved much meaning for these situations. This probability value of about 0.5 means that we should expect that we should easily be able to find new information that further resolves the probability value of each of these events into a value that is significantly higher or lower. On the other hand, if we have another set of situations for which we have resolved a probability 0.99999 of the event of occurring, then we should expect that only for very few of these situations will we be able to resolve the probability of the event to a much lower value. This is because each event is going to happen or it is not (and there is no such thing as chance in the world). If our information is complete, then we will be able to resolve each potential event into a certainty of either happening or not happening, because, other than things like the count of a Geiger counter, the only fixed (unconditional or propensity) probability values in this world are 1 and 0. And in another sense we can say that except for the probability values 1 and 0, valid probability values are never wrong. When we discover some new meaningful information then it is better to say that our old probability is superseded rather than say it is incorrect, since it is still correct given the previous information.

We might say that robust probability values are those values not likely to be updated with new information. Robust probability values are those values that seem like propensity values. The system of tossing a fair coin or a die is a special system that is so constructed, to give us robust probability values. In throwing a coin or a die, it is easy to ensure the outcome to be sufficiently sensitive to slight variations in how the die is thrown, so as to avoid the possibility of new information.

The role of randomness is to prevent the revelation of new information. In the case of bias built into the die or a coin, then the bias will reveal itself in the non-random pattern. The objective of randomness is to ensure that we can say that each outcome is independent of the previous outcome or that each outcome provides no additional information about the future outcomes (other than we can get a heads and tails for a coin and 1 to 6 for a die). A non-random sequence is one that predicts itself. When we can see the pattern is not random, then we will become aware of the bias and therefore we will want to update our probability value in accordance with our new information.

Suppose, for example, a die was always put in the cup with the six upwards, then we might find that we need to adjust our relative preferences very slightly, but after we have adjusted the relative preferences, we will again say that we have no other information about the outcome or that the outcome is otherwise random. The objective of randomness is to eliminate all patterns because patterns provide additional information.

When we use a computer algorithm to give us randomness, then the objective of the algorithm is merely to hide the pattern from accidental or deliberate discovery by making it sufficiently complex. The simpler the pattern, the more likely it is that we will deliberately or inadvertently discover the pattern. Randomness is what is left after we use all available information or rather randomness is no other information. By ensuring that there is sufficient randomness in the throw of a die we ensure that we can repeat the event without ever feeling that we are introducing new information that updates our old probability value. Therefore, we can say that we will have the same relative preference each time we throw the die, rather than only the first time we throw the die.

Perhaps it is because of our experience with robust probability values that we normally associate with tossing a coin or a die, that we forget that probability values are dynamic values that depend on the information. Information probability values are not necessarily robust values. The way we use information can seem a little strange. If we are expecting a cheque in the mail then when we go to pick the mail up, we say 'please let it be there' as if the absence of information means the outcome is not yet determined. Information probability allows us to say that the probability of being dealt an ace of hearts is 1/52 regardless that the ace of hearts has already been dealt which seems a little strange. Instead of saying that such probability values do not make any sense, most people would define a context to make sense of such probability values. However, because there is a possibility of the same sort of confusion reappearing in complex situations, it helps to understand the source of the confusion.

Logic requires that the elements of set A to remain in set A and even though such a requirement seems so ultra-trivial that it could never lead to a practical problem, we can be confused precisely because we forget that probability values are dynamic values that depend on the information at hand. That is, we become confused because propensity thinking attempts to attach probability values to events rather than to the information we have about the events. Only propensity values are fixed and never change in the light of new information and when we mistake probabilities for propensities then we isolate the probability values from the information.

Consider 'the probability that all perfect numbers are even, is 5%'. How can that make sense given that the outcome is already determined? Propensity thinking would argue that either all perfect numbers are even or they are not, and there cannot be a 5% chance that all perfect numbers are even (A perfect number is an integer number that is equal to the sum of all its divisors not including the number itself. The number 6 is a perfect number because it is divisible by 1, 2 and 3 and 1 + 2 + 3 = 6. The first three perfect numbers 6, 28 and 496 are even, and in fact, all known perfect numbers are even numbers).

When we argue that 'the probability that all perfect numbers are even is 5% does not make any sense because the conjecture is either true or false', then we are trying to permanently associate a probability value with a conjecture (like we permanently associate a probability value with the outcome from a future coin toss), when the probability value should only be permanently associated with the information we have about the conjecture. We should be interested in the set of equivalent conjectures. We can have equivalent conjectures because equivalence means equivalence in terms of information we have about the conjecture.

An event may have a probability of 50% at one moment and then some time later it seems to relocate itself to the set of events that have a probability of 10% and nothing about the event has changed. Person A may observe some trend about a population and person B may observe some other trend. If we collect all these probabilities, then we will have a lot of 'contradictory' probabilities. But, if we understand that probabilities are only permanently associated with information rather than events, then such contradictory probabilities are not contradictory.

Information probability requires that we make the best use of the information we have at hand. We update our information probability values with new information, whereas we correct our propensity values with new information. When we have a probability value that is difficult to update, then we have a robust probability value that seems to be attached to the event rather than to the information, and although that seems like the sort of probability we might want, it is also the sort of probability value that confuses our understanding. When we do not have all the information, it also makes sense to make the best use of the information we have because as Hume argues: "Though there be no such thing as *Chance* in the world; our ignorance of the real cause of any event has the same influence on the understanding, and begets a like species of belief or opinion"

Although the system of tossing a coin or a die is a special system that is so constructed that we can have repetition without being too concerned about the possibility of getting new information, even these probability values are not propensity values. The probability values associated with the next toss of the coin or die, are not in fact permanently associated with the next toss of the coin or die but rather they are permanently associated with the information we have about the next toss of the coin or die. They are robust probability values because it seems difficult to access all the information. Nevertheless, if the information changes, the probability value can change, even though we are referring to the same situation and nothing about the situation has changed. Probabilities we get from tossing a coin are robust but nevertheless, they are conditional probability values and conditional probability values are dynamic values. If we knew everything about how we tossed a coin, then we would know the outcome.

Suppose that we are drawing a red marble from a bag of red and black marbles when we do not know the proportion of marbles in the bag. Perhaps nobody would have a problem with saying the probability that 'our guess' (whatever it is) will have a 50% chance of being right (since if we have no further information we could use a coin toss to do our guessing for us), though we might hesitate to say that the probability of 'drawing a red marble is 50%'. We would be happy to say that the probability of our guess is 50% because when we describe the probability value like that, then it is clear that the probability value we have in mind is not attached to the event. But in fact, probability values should always refer to the information we have about the events and not to the events. Hence, we can also say that the first draw (given that there are red and black marbles) will have a 50% probability of producing a red marble, even though when we continue to draw more and more marbles we find only 10% of the marbles are red.

Because these probability values have nothing to do with what is in the bag, these probability values are said to come from symmetry. I prefer to say that they come from the best use of information. Some people argue that there is no such kind of probability (when we do not know the proportion of red and black marbles). Some people say that if we know that we have red, white, and blue marbles, then because we have the choice of whether to say that we have three categories or two categories (a white category and a non-white category), the probability values do not make any sense. But, when we want to combine two categories like this, then we need some additional information about the categories, otherwise if we had 100,000 categories (using numbers 1 to 100,000), then we could divide the categories into category '1' and the 'not-1' category.

In fact, the probability of drawing a red marble when we do not know the proportion of red to black marbles is an ordinary probability value with a little extra confusion. The information we have determines what set of situations are equivalent situations and in this case we have no information about the fraction of red marbles in the bag of marbles. If we knew the fraction of red and black marbles, then we would have some information that is permanently associated with the bag. We need to take another approach in order to make best use of the information. We can say that the 50%-probability (given there are two colours of marbles) is a true probability because when we put it back into the set of our other equivalent two-outcome probability events (in our mind say), which we can do providing we do not have any other information to indicate that the particular situation does not belong in this set, then we will see that it does indeed have a probability of 50%. The fact that 10% is a better estimate of the probability than 50% does not indicate that 50% is not a true probability any more than seeing the colour of a marble after it is drawn indicates that the probability was not a true probability. Probability values are about the information we have about the event, and the information we have about the event tells us which events are equivalent events.

In part II of section IV of his *Enquiry*, Hume says: "It is impossible, therefore, that any arguments from experience can prove this resemblance of the past to the future; since all these arguments are founded on the supposition of that resemblance. Let the course of things be allowed hitherto ever so regular; that alone, without some new argument or inference, proves not that, for the future, it will continue so." How can the past pattern tell us anything about the future pattern? In a sense, it doesn't tell us anything about the future pattern but it does tell us something about the probable future pattern because as Hume says: "Though there be no such thing as *Chance* in the world; our ignorance of the real cause of any event has the same influence on the understanding, and begets a like species of belief or opinion"

Hume, Russell and Popper argue that 'if we do not know that the pattern is the same, then we want the pattern to be probably the same and in this case it is necessary to assume that the past provides a probable picture of the future but this we can only do by assuming that the past represents the future which is the point in question'. However, this argument is based on the idea that such probable arguments can only proceed on the basis of similarity of one situation with another.

The reason why we say that we can make an inductive inference to the near future is because we are looking to find the *special* point of discontinuity — the point of break in the pattern. Although it might be illogical to say the future (or rather the total future) will resemble the past, we are still looking for that special point. So, providing we are not unduly interfering with the pattern, and we have no other information, we can always say that it is unlikely that now is that special time. It is a mistake to say that because we know nothing about the pattern (except for the past pattern); we cannot have probable information about the future pattern. It is because we know nothing about the pattern that we can be certain that our position is a random in its relationship to the point of discontinuity. If we are not past the end of the pattern, then we can say that we are probably near the middle of the pattern, which is an argument that has nothing to do with sampling. The fact that for all we know we might be on the edge of the discontinuity (i.e. the break in the pattern), does not mean that induction has not got a probable status. This additional information will only supersede (not invalidate) our probability and it will only do this if we have such information, not if we imagine we have such information.

Popper said that if induction is probable then the probability must be tautological. This means that if a principle of induction is true, then it must be true regardless of how the pattern was generated. In fact, from the point of view of inductive information (which is the only information we are interested in), we can say that all patterns are equivalent if they have the same number of confirming instances and that is the only information we have about the patterns.

Suppose we are trying to guess the size of a number given only that the number is larger than zero, when the only feedback we get for each guess is 'the number is larger than the guess' or 'the guess has busted the number' (if our guess has overshot the number). In this analogy, the location defined by the number is the location where the pattern runs out, which is to say the location of the discontinuity.

Suppose we guess 68 and we are told that the number is larger, we would immediately assume that the actual number has at least a 50% chance of being equal to or larger than 2*68 = 136. Now this is a form of induction because we think that we have some information about the number (which is the future) when we only have experience with the past. This can only be true if we have a reason for trusting some form of type 1 inductive reasoning and that we can use the past to see into the future. The inductive prediction does not refer to this particular pattern but only a pattern with length L because the probability of an event is associated with the information we have about the event and not about the event itself. In other words, even though inductive reasoning cannot see into the future, it can see into the probable future.

In the case of guessing a number, our reasoning says that we know that our guess is somewhere between the beginning and the number, and that (assuming we have not busted) there is no reason to think that our guess will have a preference for either the beginning or the end of the pattern. If there is no inherent preference for the beginning or the end, then the best guess should be in the middle. We are in fact saying that our guess is random in its relation to the actual point of discontinuity, which it must be if we do not know anything about the pattern (and the method of generating the pattern does not know anything about how we guess). This is also true as we increase the number — the only information we have is that we have not busted and therefore the only logical conclusion we can make is that we are probably in the middle of the pattern.

We should say *not quite* the middle because we have not specified an upper limit and therefore there are reasons to suspect that our guess might have a preference for the beginning of the pattern. What we should say is that if we guess a number P and have not busted then there is at least a 50% chance of the number being 2*P or larger. Also, the probability that we will bust if we increase the number by F is less than or equal to: F/(P+F). For this number guessing exercise there must be no additional information about the target number.

We already understand this kind of thinking because we all practice it. Hume forgot that probability values refer to the information we have about the situation and therefore the set of equivalent situations rather than the actual situation. We can't use the past to see in the future but we can use the past to see into the probable future.

The operation of a silo dumping sand (or molecules of water) is analogous to our number guessing experiment, and provides a Galileo thought experiment with inescapable conclusions. Suppose, for example, we have a very large silo which is storing a finite amount of sand. Because we do not want any additional information the rate of dump must be constant regardless of the amount of sand in the silo. We press the start button and the silo empties onto a conveyor, and some time later (say a few weeks later) we press the stop button and the silo instantly stops. Now we can be certain that when we press the start button again, that the immediate future will resemble the past and the silo will provide some more grains of sand. Clearly our experience does not have access to the full domain and yet we can say that it is extremely unlikely that we have found the point of discontinuity. And even when we stipulated that it was impossible that the past was going to resemble the future (because we specified that the silo held only a finite amount of sand), the past still *probably* predicts the near future.

If we have dumped say 10^{12} grains of sand when we turned the machine off, then we have *at least* a 50% chance of dumping another 10^{12} grains of sand when we turn the machine back on.

This probability does not make any predictions about the pattern, except to say that if our current status is 'such and such', it is unlikely that the point of discontinuity is very close to us. The probability works because there is no information connecting the point where we have stopped and the point of discontinuity, and therefore no preference/bias for either the beginning or the end of the pattern.

While it is easy to see that we will have at least a 50% chance of getting the same amount of sand as we have already dumped (on the first attempt), it is more difficult to appreciate that we can use this same argument for each additional grain of sand that we dump. To say that, if we have dumped 10^{12} grains of sand that we can then dump one more grain of sand to give us some additional information about the silo seems paradoxical, particularly when the thing we are most certain of is that we have moved one more grain of sand closer to busting.

Probable information must be conveyed by each additional grain of sand since if we have reached 10^{12} grains of sand one at a time or in one go, in the end we must have exactly the same information. It is true that we have moved one more grain of sand closer to the actual discontinuity, the probability value should not be associated with the actual silo (because probability values are not propensity values) but it should be associated with the information we have about the silo. Although we have moved closer to the actual point of discontinuity we have moved further from the probable location of the discontinuity.

So, we can make inductive inferences into the near future (if we are not interfering with the pattern and we have no other information about the pattern) and the number of prior instances of 'X' does give some information about the number of future instances we can expect, and the probability of the next future instance of 'X'. And this principle of induction is not empirical — it is in fact tautological and will be true in every possible universe as Popper required. Induction works for the tautological reason that 'this instance is probably not a special instance'. On page 384 (Appendix *VII), of his *Logic of Scientific Discovery*, Popper writes:

> We may sum up our criticism of (+) as follows. Some people believe that, for purely logical reasons, the probability that the next thing we meet will be red increases in general with the number of red things seen in the past. But this is a belief in magic — in the magic of human language. For 'red' is merely a predicate; and there will always be predicates *A* and *B* which both apply to all the things so far observed, but lead to incompatible probabilistic predictions with respect to the next thing. These predicates may not occur in ordinary languages, but they can always be constructed.

Suppose we repeated our experiment with our silo and noted that only red things came out of the silo. If we slam the silo closed at some instant, we can be sure that the next thing will be red — because logic allows us to see into the future in a probable sense. If the number of previous events is small, it might be possible to graph the variations in the red and make a non-red prediction for the near future, but non-red descriptions would not be equally confirmed. If we call data that has been used to construct the description, ad hoc data and the data that has not been used to construct the description redundant data, then it should be clear that only the *redundant* data can be used to confirm our description, and the more ad hoc data we need to generate our description, the less redundant data we will have to confirm our description so there is nothing paradoxical here.

In section 1.2, it was noted that induction can fail because we do not determine all the variables. Although this was an attempt at empirical induction, empirical induction when we do not determine all the variables is induction by enumeration. Consider, for example, the previous example of trees that 'fruit in December' where we happen to be located in Australia. How do we know that 'December' is not an arbitrary descriptor, given that we can find other descriptors such as 'summer'? This particular problem and many like it suggests that induction only works if we are naive. However, all probability values do in fact need to be appropriately naive because 'appropriately naive' is just a way of saying no additional information. If we did not identify season as a potential variable that describes the pattern, then the probability would be correctly calculated, even though the climate theorists would understand the confidence of their colleagues to be inappropriate. If we identify 'seasons' or 'December in Australia' as another independent variable that describes the pattern, then our induction would be superseded because both descriptions fit the pattern and both descriptions are equally confirmed by the redundant data. We might say that even though our old inductive probability was a true probability, in the light of the new information it is now a superseded probability.

Induction by enumeration is not interested in new information because it has no interest in the factors whereas empirical induction invokes our understanding of variables. Charles Darwin suggested that blue eyed cats are deaf. Clearly this is purely induction by enumeration because a scientists from Darwin's time would not be able to anticipate a link between blue eyes and deafness. Sometime later Darwin noted that 'white cats with blue eyes are almost always deaf'. Again even if the colour of the cat is a true empirical factor, for Darwin this new factor was not necessarily a true empirical factor in the sense of empirical induction. For the colour of the cat to be a true empirical factor in the sense of empirical induction, Darwin would need only note that blue eyes relates to a lack of pigmentation and the white colour relates to a lack of pigmentation and somehow surmise that a lack of pigmentation might be the factor that correlates with deafness. Such empirical thinking would give him new information even if he did not understand exactly how pigmentation could be related to deafness. When we have no understanding of the variables, then it is not possible to notice anything different about the next instance.

If all swans have been white then, the next swan will probably be white. If we have some understanding of the situation then we may notice something different about the next instance that might make the next instance non-typical and we may want to use our empirical thinking to provide an alternative conclusion. If we were simply noting how many swans have been white, then we have no other information that can interfere with our inductive thinking— we just say 'all swans have been white'. In this case, the only information we have is the number of swans that have been white.

In other cases where we think we have some understanding of the variables that might affect the outcome (such as the time when trees fruit), then we will need to use Mill's method. Mill's method assumes that we have got some principle of repeatability to fall back on — so that we do not use induction by enumeration to make predictions, but instead we determine the variables and assume that repeatability will hold true. However, we need induction by enumeration to say that the principle of repeatability will always hold true. And although induction by enumeration will not allow us to say that the principle of repeatability will always hold true, the validity of type 1 induction means that we have reason to trust the principle of repeatability for the near instances, with near instances being a number of instances that is small compared with the number of confirmations.

2.2 Type 2 Induction

On page 154 of his *Conjectures and Refutations*, Popper compares laws of science with the laws of mathematics and he discusses three views on scientific theories. The first view he discusses is what he calls the Galilean view that the scientist can establish the truth of his theories beyond all reasonable doubt. The second view is the 'instrumentalist' view. The third view he discusses is his view:

> This 'third view' is not very startling or even surprising, I think. It preserves the Galilean doctrine that the scientist aims at a true description of the world, or of some of its aspects, and at a true explanation of observable facts; and it combines this doctrine with the non-Galilean view that though this remains the aim of the scientist, he can never know for certain whether his findings are true, although he may sometimes establish with reasonable certainty that a theory is false. One may formulate this 'third view' of scientific theories briefly by saying that they are *genuine conjectures* — highly informative guesses about the world which although not verifiable (i.e. capable of being shown to be true) can be submitted to critical tests. They are serious attempts to discover the truth. In this respect scientific hypotheses are exactly like Goldbach's famous conjecture in the theory of numbers. Goldbach thought that it might possibly be true; and it may well be true in fact, even though *we do not know, and may perhaps never know, whether itis true or not.*

Popper claims that the probability of a scientific theory being true is zero. If Popper truly believes that the Goldbach conjecture is exactly like scientific hypotheses then he should be happy to claim that 'we cannot even say that the Goldbach conjecture is probably true'.

The Goldbach conjecture (conjectured by Goldbach in a letter to Leonhard Euler in 1742) claims that every even integer greater than two can be written as the sum of two, not necessarily distinct primes. Confirmations of the Goldbach conjecture include: 4 = 2 + 2, 6 = 3 + 3, 8 = 3 + 5, 10 = 7 + 3, etc. Although there is no known exception to this conjecture, like many other conjectures, there is no known proof of this conjecture either.

As the size of the even numbers increase, the number of available Goldbach pairs (GP's) increases in a haphazard pattern — for the number 1050 there are 61 GP's (and we only need one to satisfy our conjecture). We could delete the primes, 11, 17, 29, 41, 59, 67, 71, etc and the Goldbach conjecture would still seem to be true, so with all the prime numbers included the conjecture should be a safer conjecture. Or in other words, if we are confident that we will have 4 GP's for even numbers larger than 100, we should be much more confident that there is at least 1 GP for even numbers larger than 100. Alternatively, from a graph of the number of pairs of primes plotted against the even number, it seems that we can make an endless number of apparently reliable inductions. For example for all even numbers 14 or larger there are two or more GP's (the even number 12 has only 1 GP). For all even numbers 70 or larger, there are three or more GP's (the even number 68 has 2 GP's). For all even numbers 100 or larger, there are 4 or more GP's (the even number 98 has 3 GP's). Therefore it seems reasonable that we should feel comfortable with the Goldbach conjecture.

The Goldbach conjecture seems to be a reliable type 2 induction, even though we have no way of knowing that it will be true. Even to say that the average spacing of prime numbers gets further apart as the size of the prime number increases is also a type 2 conjecture, if we are conjecturing that this trend will continue to infinity. The fact that we use induction like this must mean that we think that confirmations count and that we can attribute some non-zero probability value to conjectured mathematical patterns.

When Popper speaks of the probability of Goldbach conjecture, he has in mind the infinite opportunities for the Goldbach conjecture failing, whereas the information probability perspective allows us to think of the Goldbach conjecture in terms of the probability of equivalent conjectures.

There are many laws of mathematics that we know are true laws because they can be proved to be true. Therefore it seems reasonable to suspect that the Goldbach conjecture and the conjecture that all perfect numbers are even are true laws. Knowing that the first 1000 even numbers can be written as the sum of two primes means more than knowing that 4 can be written as the sum of two primes. We can be a little sceptical. The prime number pattern isn't a smooth curve suggesting that it may be different from other mathematical patterns. However, Popper claims that the probability of scientific laws is zero even though he likens scientific laws to mathematical laws and we know that that there are many mathematical laws that are true for all instances. Using the equivalent information perspective to assert a zero probability is being certain not sceptical. On p 45 of *Conjectures and Refutations*, in a discussion of how we obtain our knowledge, Popper in fact argues that induction is irrational:

> Hume, I felt, had never accepted the full force of his own logical analysis. Having refuted the logical idea of induction he was faced with the following problem: how do we actually obtain our knowledge, as a matter of psychological fact, if induction is logically invalid and rationally unjustifiable? There are two possible answers: (1) We obtain our knowledge by a non-inductive

> procedure. This answer would have allowed Hume to retain a form of rationalism. (2) We obtain our knowledge by repetition and induction, and therefore by a logically invalid and rationally unjustifiable procedure, so that all apparent knowledge is merely a kind of belief — belief based on habit. This answer would imply that even scientific knowledge is irrational, so that rationalism is absurd, and must be given up. (I shall not discuss here the age old attempts, now again fashionable, to get out of the difficulty by asserting that though induction is of course logically invalid if we mean by 'logic' the same as 'deductive logic', it is not irrational by its own standards, and as inductive logic admits; as may be seen from the fact that every reasonable man applies it as a matter of fact. Against this, it was Hume's great achievement to break this uncritical identification of the question of fact — *quid facti?* — and the question of justification or validity — *quid juris?*

If we truly do not believe in induction, then we should not make conjectures on the basis of confirming instances because confirming instances would make no sense. But there are many conjectures, such as the Goldbach conjecture, that are conjectured because there are many instances for which they are known to be true. Mathematics is full of patterns that can be discovered by observation and many such patterns can be proved to be true.

Empirical patterns are different. Regarding empirical laws, we do not know if there are any universal laws because we do not get to prove the laws of science.

2.3 Empirical Induction

Mill's solution to Hume's main problem left a few loose ends because many of the laws of science are written in the language of mathematics and mathematical laws cannot be validated by Mill's principle of repeatability. Newton's F = M*A says that this expression will be true for all values of F and M and the principle of repeatability does not help us to say that it will be true for all values, because we can never test it for the infinite combinations of possible values. In order to say that some analogue law will be true within the domain of tested values or probably true outside of the domain of tested values, we need to assume something like a principle of simplicity.

The following citation has been taken from pages 120, 123 &124 of Popper's *Logic of Scientific Discovery.* Popper discusses the problem of induction in regard to mathematical curves within the context of his discussion of simplicity. Popper's reference to Weyl is a reference to p 156 of Herman Weyl's book *Philosophy of Mathematics and Natural Science*, Princton, 1949:

> There seems to be little agreement as to the importance of the so-called 'problem of simplicity'. Weyl said, not long ago, that 'the problem of simplicity is of central importance for the epistemology of natural sciences'. Yet it seems that interest in the problem has lately declined; perhaps because, especially after Weyl's penetrating analysis, there seemed to be so little chance of solving it.
>
> Until quite recently the idea of simplicity has been used uncritically, as though it were obvious what simplicity is, and why it should be valuable. Not a few philosophers of

science have given the concept of simplicity a place of crucial importance in their theories, without even noticing the difficulties to which it gives rise. For example, the followers of Mach, Kirchhoff, and Avenarius have tried to replace the idea of a causal explanation by that of the 'simplest description'. Without the adjective 'simplest' or a similar word this doctrine would say nothing. As it is supposed to explain why we prefer a description of the world with the help of theories to one with the help of singular statements, it seems to presuppose that theories are simpler that singular statements. Yet few have ever attempted to explain why they should be simpler, or what is meant, more precisely, by simplicity...

The epistemological idea of simplicity plays a special part in theories of inductive logic, for example in connection with the problem of the 'simplest curve'. Believers in inductive logic assume that we arrive at natural laws by generalisation from particular observations. If we think of the various results in a series of observations as points plotted in a co-ordinate system, then the graphic representation of the law will be a curve passing through all these points. But through a finite number of points we can always draw an unlimited number of curves of the most diverse form. Since therefore the law is not uniquely determined by the observations, inductive logic is confronted with the problem of deciding which curve, among all these possible curves is to be chosen.

The usual answer is, 'choose the simplest curve'. Wittgenstein, for example, says: 'The process of induction consists in assuming the *simplest* law that can be made to harmonize with our experience. In choosing the simplest law, it is usually tacitly assumed that a linear function, say, is simpler that a quadratic one, a circle simpler than an ellipse, etc. But no reasons are given either for choosing this particular hierarchy of simplicities in preference to any other, or for believing that 'simple' laws have advantages over the less simple — apart from aesthetic and practical ones. Schlick and Feigl mention an unpublished paper of Natkin who, according to Schlick's account, proposed to call one curve simpler than another if its average curvature is smaller; or, according to Feigl's account, if it deviates less from the straight line...

Weyl discusses and rejects a very interesting attempt to base simplicity on probability. "Assume, for example, that twenty co-ordinated pairs of values(x,y) of the same function, y = f(x) lie (within the expected accuracy) on a straight line, when plotted on square graph paper. We shall then conjecture that we are faced here with a rigorous natural law, and that y depends linearly on x. And we shall conjecture this because of the *simplicity* of the straight line, or because it would be so *extremely improbable* that just these twenty pairs of arbitrary chosen observations should lie very nearly on a straight line, had the law in question been a different one. If now we use the straight line for interpolation and extrapolation, we obtain predictions which go beyond what the observations tell us. However, this analysis is open to criticism. It will always be possible to define all kinds of mathematical functions which...will be satisfied by the twenty observations; and some of these functions will deviate considerably from the straight line. And for every single one of these we may claim that it would be *extremely improbable* that the twenty observations should lie just on this curve, unless it represented the true law. It is thus essential, after all, the function, or rather the class of functions, should be offered to us,*a priori*, by mathematics because of its mathematical simplicity. It should be noted that this class of functions must not depend upon as many parameters as the number of observations to be satisfied." Weyl's

> remark that 'the class of function should be offered to us *a priori*, by mathematics, because of it mathematical simplicity', and his reference to the number of parameters agree with my view (to be developed in section 43). But Weyl does not say what 'mathematical simplicity' is; and above all, he does not say what *logical or epistemological advantages* the simpler law is supposed to possess, compared with one that is more complex...

Concerning our analogue laws, perhaps the most fundamental assumption we would want to make is that there are few or no special points in our mathematical curves that we use to describe the laws. The more special points we have the less simple the law, because each point needs its own bit of information. If we had a hand drawn curve and then we would need an infinite number of data points to describe the curve. If we do not reject such special points, then we might expect that at some state, say at distance $x = 2.13...\ m$ between two bodies, that some fixed parameter in the law of gravity will undergo a change.

If we have any finite number of '*n*' points (let's say 20 points) that lie on any hand-drawn function, then we can always find a curve that fits the n points. We can do this because techniques for fitting, say a polynomial curve to multiple points effectively search the infinite number of polynomial solutions and come up with a perfect fit to the data. If we only need say three coefficients to provide a fit though all twenty points then we would feel that we have found a good approximation to the curve. We may not be able to extrapolate large distances, but we can have some confidence that every interpolated point will lie approximately on the curve. This is because we only have 3 ad hoc data, which leaves 17 redundant data to confirm our description.

This curve fitting exercise (when we have lots of redundant data) is analogous to the situation of having a simple expression for a law of gravitation (in conjunction with some simple laws of dynamics) that explains falling bodies on earth, attraction of bodies on the earth, the elliptical orbits of moons and planets, and the rise and fall of the tides, etc. The future and the past description of the position, velocity and acceleration of all planets, stars, moons, comets and meteors throughout the entire universe can apparently be compacted into the simple laws of dynamics and the law of gravitation (and some initial conditions). In a letter to Leibniz 16 Oct 1693 (cited in Westfall p 509), Newton writes:

> For since celestial motions are more regular than if they arose from vortices and observe other laws, so much so that vortices contribute not to the regulation but to the disturbance of the motions of planets and comets; and since all phenomena of the heavens and of the sea follow precisely, so far as I am aware, from nothing but gravity acting in accordance with the laws described by me; and since nature is very simple, I have myself concluded that all other causes are to be rejected and that the heavens are to be stripped as far as may be of all matter, lest the motions of planets and comets be hindered or rendered irregular. But if, meanwhile, someone explains gravity along with all its laws by the action of some subtle matter, and shows that the motion of planets and comets will not be disturbed by this matter, I shall be far from objecting.

In the above, Newton is defending his theory against claims that a true explanation must have vortices of some subtle matter (so as to be a true mechanical cause) to carry the planets in their orbits rather than magical laws of gravity. Late in his life, Newton comments on the general philosophical problem of conjecturing solutions to problems (Westfall p 642-643):

> Could all the phenomena of nature be deduced from only three or four general suppositions there might be great reason to allow those suppositions to be true: but if for

> explaining every new Phenomenon you make a new Hypothesis ... your Philosophy will be nothing else than a system of Hypotheses. And what certainty can there be in a Philosophy which consists in as many Hypotheses as there are Phenomena to be explained.

Whereas, Occam's razor relates reliability to the fewest hypotheses Newton relates reliability the idea of having few hypotheses relative to the number of phenomena. Newton's statement is about ad hoc data, redundant data and simplicity. Given Newton's familiarity with polynomials, it seems that Newton had little choice but to understand such concepts. With a minor adjustment to Newton's words above we could make Newton say: 'What certainty can there be to a curve fit that consists in as many coefficients as there are data to be explained?'.

Mathematical laws are simple. We do not describe the line y = x as the group of points that includes the point at the position (56, 56) and another at (1, 1) and another at (2.2, 2.2) etc, because this is cumbersome and we want efficient and simple. Instead, we might say that the group of points lies on the straight line that passes through (0,0) and has a slope of 1:1.

When we describe the points as lying on a straight line with slope 1:1, then we are saying that we have found the constant in the pattern. If the slope of the line is increasing at a constant rate (such as $Y = X^2$), then when we have found the rate of increase then we have found the *constant* in the pattern. The amount of ad hoc data required to describe the pattern is a measure of the complexity of the pattern. The remaining redundant data can then be used to confirm the pattern. The more complex the pattern, the less data we will have for confirmation. Thus 'simplicity', 'confirmation', 'ad hoc' and a 'constant pattern' are similar concepts.

One difficulty with the idea that we can confirm a simple mathematical description with redundant data is that we can confirm a known ad hoc solution (say a cubic spline) adjacent to the ad hoc data with an infinite number of new data points. However, this is possible because empirical data often fits on a smooth curve and a sufficiently small portion of a smooth curve will seem straight *when our data is approximate*. This means we can provide a multitude of interpolative and slightly extrapolative confirmations for a description that is not the true description because approximate data means that there will be an ad hoc region. The concept of a smooth curve is similar to the concept of 'no special points'.

Even if there are no special points, it must be the case that 'the trend' and our description of the trend are not necessarily the same thing, because when our data is approximate there is no unique solution. It must be possible that we can miss some variables, either because we have *a wrong view of the way things are* or because *measurement error* does not allow us to pick up subtle variables. Gravity seems constant as we climb the stairs. If we happen to be very observant mountain climbers, then the simplest description might be that gravity varies linearly with height. But if we are space travellers then, we will see that gravity varies as the inverse of the square of the distance. But even if we know that $y = a x^2$ describes some data over a very wide range of redundant data and even though we expect the law to remain the same, we have no way of knowing that the law is not $y = ax^2 + bx^3$ (say) because we can always make 'b' sufficiently small so that the effect of this additional term is not visible.

It is also true that Mill's solution to Hume's problem only works if we think we can guess which variables are relevant and which are not. The car mechanic has the advantage of knowing that changing the windscreen wipers will not help the car run better but there are no methods (other than the use of our imagination and trial and error) for determining all the variables, so our laws will sometimes fail. Although there are no guarantees for whatever reason it does seem that modern science often has a handle on which variables are relevant and which are not. That Newton saw the need to experiment with pendulums, which were made of gold, silver, lead, glass, wheat, etc (see Westfall p 422), now seems a little strange.

There may be many other loose ends. The fact that we can discover very good approximations to the laws using simple descriptions seems to mean that laws are simple over their useful domain. This general confidence in laws, must add to our confidence in specific laws. And we have no other choice than to assume simplicity because we at least have a type 1 inference on our side whereas a theory that includes specific unconfirmed complexity includes a wild guess.

Chapter 3 POPPER'S PHILOSOPHY OF SCIENCE

Spokesmen for evolution use Popper's testability criterion to condemn creationism. On page 11 of *The Logic of Scientific Discovery (LOSD)*, Popper explains that his testability criteria separates science from non-science, and on page 20, Popper explains that his testability criterion solved the problem of induction:

> LOSD P10:
>
> The problem of finding a criterion which would enable us to distinguish between the empirical sciences on the one hand, and mathematics and logic as well as 'metaphysical' systems on the other, I call the problem of demarcation.
>
> This problem was known to Hume who attempted to solve it. With Kant it became the central problem of the theory of knowledge. If, following Kant, we call the problem of induction 'Hume's problem', we might call the problem of demarcation 'Kant's problem'.
>
> Of these two problems — the source of nearly all the other problems of the theory of knowledge — the problem of demarcation is, I think, the more fundamental. Indeed, the main reason why epistemologists with empiricist leanings tend to pin their faith to the 'method of induction' seems to be their belief that this method alone can provide a suitable criterion of demarcation. This applies especially to those empiricists who follow the flag of 'positivism'.
>
> LOSD P20:
>
> The proposed criterion of demarcation also leads us to a solution of Hume's problem of induction — of the problem

of validity of natural laws. The root of his problem is the apparent contradiction between what may be called 'the fundamental thesis of empiricism' — the thesis that experience alone can decide upon the truth or falsity of scientific statements — and Hume's realization of the inadmissibility of inductive arguments. This contradiction arises only if it is assumed that all empirical scientific statements must be 'conclusively decidable', i.e. that their verification and their falsification must both in principle be possible. If we renounce this requirement and admit as empirical also statements which are decidable in one sense only — unilaterally decidable and, more especially, falsifiable — and which may be tested by systematic attempts to falsify them, the contradiction disappears: the method of falsification presupposes no inductive inference, but only the tautological transformations of deductive logic whose validity is not in dispute.

3.1 Testability and Induction

On page 10 of the *Logic of Scientific Discovery,* Popper argues that we can escape the problem of induction if we accept that our *knowledge is conjectural* — that we can never prove a theory true or even probably true but that we can prove a theory to be false:

> Next we seek a decision as regards these (and other) derived statements by comparing them with the results of practical applications and experiments. If this decision is positive, that is, if the singular conclusions turn out to be acceptable, or *verified*, then the theory has, for the time being, passed its test: we have found no reason to discard it. But if the decision is negative, or in other words, if the conclusions have been *falsified*, then their falsification also falsifies the theory from which they were logically deduced.
>
> It should be noticed that a positive decision can only temporarily support the theory, for subsequent negative decisions may always overthrow it. So long as a theory withstands detailed and severe tests and is not superseded by another theory in the course of scientific progress, we may say that it has 'proved its mettle' or that it is '*corroborated*' by past experience.
>
> Nothing resembling inductive logic appears in the procedure here outlined. I never assume that we can argue from the truth of singular statements to the truth of theories. I never assume that by force of 'verified' conclusions, theories can be established as 'true', or even merely 'probable'. In this book I intend to give a more detailed analysis of the methods of deductive testing. And I shall attempt to show that, within the framework

> of this analysis, all the problems can be dealt with that are usually called '*epistemological*'. Those problems, more especially, to which inductive logic gives rise, can be eliminated without creating new ones in their place.

If there is no such thing as induction, then there is no such thing as confirmation of a pattern. In the place of the word 'confirmation', Popper uses words like 'corroboration' and 'proved its mettle'. He effectively defines 'corroboration' to mean 'currently works' but on pages 274-275 of *The Logic of Scientific Discovery*, Popper makes it clear that such a term does not imply any kind of reliability, probability, truth, or any logical reason to trust the theory:

> This same timelessness we also attach to the concepts 'true' and 'false', in agreement with common usage. It is not common usage to say of a statement that it was perfectly true yesterday but has become false today. If yesterday we appraised a statement as true which today we appraise as false, then we implicitly assert today that *we were mistaken yesterday;* that the statement was false even yesterday — timelessly false — but that we erroneously 'took it for true'.
>
> Here one can see very clearly the difference between truth and corroboration. The appraisal of a statement as corroborated or as not corroborated is also a logical appraisal and therefore also timeless; for it asserts that a certain logical relation holds between a theoretical system and some system of accepted basic statements. But we can never simply say of a statement that it is as such, or in itself, 'corroborated' (in the way in which we may say that it is 'true'). We can only say that it is *corroborated with respect to some system of basic statements* — a system accepted up to a particular point in time. 'The corroboration which a theory has received up to yesterday' is *logically not identical* with 'the corroboration — a subscript characterizing the system of basic statements to

> which the corroboration relates' (for example, by the date of its acceptance).
>
> Corroboration is therefore not a 'truth value'; that is, it cannot be placed on a par with the concepts 'true' and 'false' (which are free from temporal subscripts); for to one and the same statement there may be any number of different corroboration values, of which indeed all can be 'correct' or 'true' at the same time. For they are values which are logically derivable from the theory and the various sets of basic statements accepted at various times.
>
> The above remarks may also help to elucidate the contrast between my views and those of the pragmatists who propose to *define 'truth' in terms of the success of a theory — and thus of its usefulness, or of its confirmation or of its corroboration.* If their intention is merely to assert that a logical appraisal of the success of a theory can be no more than an appraisal of its corroboration, I agree. But I think that it would be far from '*useful*' to identify the concept of corroboration with that of truth.[*3] This is also avoided in ordinary language. For one might say of a theory that it has hardly been corroborated at all so far, or that it is still uncorroborated. But we should not normally say of a theory that it is hardly true at all so far, or that it is still false.
>
> Popper's Footnote *3 reads: Thus if we were to define 'true' as 'useful' (as suggested by some pragmatists), or else as 'successful' or 'confirmed' or 'corroborated', we should only have to introduce a new 'absolute' and 'timeless' concept to play the role of 'truth'.

Aristotle said something similar to Popper in his On *Interpretations*, part 1 section 19:

> Let me illustrate. A sea-battle must either take place to-morrow or not, but it is not necessary that it should take place to-morrow, neither is it necessary that it should

> not take place, yet it is necessary that it either should or should not take place to-morrow. Since propositions correspond with facts, it is evident that when in future events there is a real alternative, and a potentiality in contrary directions, the corresponding affirmation and denial have the same character.
>
> This is the case with regard to that which is not always existent or not always nonexistent. One of the two propositions in such instances must be true and the other false, but we cannot say determinately that this or that is false, but must leave the alternative undecided. One may indeed be more likely to be true than the other, but it cannot be either actually true or actually false. It is therefore plain that it is not necessary that of an affirmation and a denial one should be true and the other false. For in the case of that which exists potentially, but not actually, the rule which applies to that which exists actually does not hold good. The case is rather as we have indicated.

Both Aristotle and Popper seem to be trying to permanently attach a truth value to a particular outcome, but probability values aren't like that. When we speak of the probability of an event, we should always mean the probability of an event given certain information. I can say that 'I probably do not hold the ace of hearts in my hand' if I have not looked in my hand. Propensity values are permanently attached to events, but probability values are attached to the information we have about the event. The nature of information is that something may be both true (highly probable) and impossible. What was perfectly true yesterday (i.e. what was highly probable yesterday) may become false (highly improbable or even preposterously impossible) today, because probability values correspond to information and information changes with time and perspective. Because of how we use information and probability, it makes sense to say 'it is highly probable that a certain proposition is true' even when it is mathematically impossible that such a proposition is true.

On page 268, *The of Logic of Scientific Discovery* Popper suggests that testing a theory in new circumstances is the best way to corroborate a theory: "When trying to appraise the degree of corroboration of a theory we may reason somewhat as follows. Its degrees of corroboration will increase with the number of its corroborating instances. Here we usually accord to the first corroborating instances far greater importance than to later ones: once a theory is well corroborated, further instances raise its degree of corroboration only very little. This rule however does not hold good if these new instances are very different from the earlier ones, that is if they corroborate the theory in a *new field of application*. In this case, they may increase the degree of corroboration very considerably." The fact that Popper prefers severe tests in new circumstances seems to imply that such tests somehow provide better information than confirmations in typical circumstances. No doubt we will all agree with the need for 'new circumstances' but the reason we agree is surely because we all agree with Newton's understanding of ad hoc data and redundant data. The classical challenge to inductive thinking namely that 'the sun might not rise tomorrow' suggests that there is no such thing as ad hoc data, whereas Popper's understanding of the need to test theories in new circumstances seems to recognise that there is such a thing as ad hoc data.

What we are really looking for is some way of providing reliability or credibility for our laws, in both everyday situations and in science. We want to be able to say that if we have 50 confirmations of some conjectured law (such as the Goldbach conjecture), then we have provided some sort of probable confirmation of the conjecture. Popper realises that we have achieved some sort of 'reliability' for our scientific theories but on pages 252 and 258 of *The Logic of Scientific*

Discovery, he argues the 'reliability' is a psychological phenomena that has no logical basis:

> Like inductive logic in general, the theory of the probability of hypotheses seems to have arisen through a confusion of psychological with logical questions. Admittedly, our subjective feelings of conviction are of different intensities, and the degree of confidence with which we await the fulfilment of a prediction and the further corroboration of a hypothesis is likely to depend, among other things, upon the way in which this hypothesis has stood up to tests so far — upon its past corroboration. But that these psychological questions do not belong to epistemology or methodology is pretty well acknowledged even by the believers in probability logic. They argue, however, that it is possible, on the basis of inductivists decisions, to ascribe degrees of probability to *the hypotheses themselves*; and further, that it is possible to reduce this concept to that of the probability of events...
>
> (5) The Psychological problem of (why do we believe that the theory so chosen will continue to be worthy of our trust?) is, I suggest, trivial: a belief or trust is always irrational, but it may be important for action....
>
> > 3 An argument which appeals to the fact that we possess knowledge or that we can learn from experience, and which concludes from this fact that knowledge or learning from experience must be possible, and further, that every theory which entails the impossibility of knowledge, or of learning from experience, must be false, may be called a 'transcendental argument'. (This is an allusion to Kant.) I believe that a transcendental argument may indeed be valid if it is used critically — against a theory which entails the impossibility of knowledge, or of learning from experience. But one must be very careful in using it. Empirical knowledge in *some sense* of the word

> 'knowledge', exists. But in another sense — for example in the sense of *certain* knowledge or of *demonstrable* knowledge — it does not. And we must not assume, uncritically, that we have 'probable' knowledge — knowledge that is probable in the sense of the calculus of probability. It is indeed my contention that we do not have probable knowledge in this sense. For I believe that what we may call 'empirical knowledge', including 'scientific knowledge', consists of guesses, and that many of these guesses are not probable (or have a probability of zero) even though they may be very well corroborated...

In that we all feel certain that our laws will not fail in the next instance, Popper must believe that our subjective psychological knowledge, is able to confer to us some valuable information that is beyond the reach of *The Logic of Scientific Discovery.* Even though Popper seems to believe in the psychological usefulness of induction, on page 258 of his *Conjectures and Refutations*, he links dogmatic thinking with a trust in induction:

> What was his error? As I have said, Kant, like almost all philosophers and epistemologists right into the twentieth century, was convinced that Newton's theory was *true.* This conviction was inescapable. Newton's theory had made the most astonishing and exact predictions, all of which had proved strikingly correct. Only ignorant men could doubt its truth...
>
> Even those who do not accept Einstein's theory of gravitation ought to admit that his was an achievement of truly epoch-making significance. For his theory established at least that Newton's theory, no matter whether true or false, was certainly *not the only possible* system of celestial mechanics that could explain the phenomena in a simple and convincing way. For the first time in more than 200 years Newton's theory had became *problematical.* It has

> become, during these two centuries, a dangerous *dogma* — a dogma of almost stupefying power.

Although it is correct to say that Newton's laws are not true universal laws, there is a sense in which we can say that Newton's laws are still true laws. What has changed is that we now know that the scope of Newton's laws is limited. There are two or three aspects to Newton's laws we need to consider — there is the principle of repeatability (that says if we do this then such and such will always result) and there is the simplifying theory aspect that brings together all relevant experience into a compact description, and there is the generalisation aspect where it is hoped that there are no other relevant factors that need to be taken into account by the theory. It could have been that Newton's laws no longer applied in the same situations. Or, we could have discovered some special points where Newton's laws failed. However, what seems to have occurred is that Newton, and the scientists that followed him had the wrong view of things and because of measurement error, they did not determine all the variables. This can be considered to be a generalisation error that Newton anticipated. On pages 320-321, of Book III of the *Principia*, Newton says regarding the *RULES OF REASONING IN PHILOSOPHY:*

> RULE IV
>
> In experimental philosophy we are to look upon propositions inferred by general induction from phenomena as accurately or very nearly true, notwithstanding any contrary hypotheses that may be imagined, till such time as other phenomena occur, by which they may be either made more accurate or liable to exceptions.
>
> This rule we must follow, that the argument of induction may not be evaded by hypotheses.

Hume's argument was about evading inductive reasoning by hypotheses. Though Newton's statement makes it clear that simplicity does not give us certainty, it seems true to say that the reason scientists were so confident that the Newtonian view of things was the correct view was because they had been programmed by simplicity and inductions derived from everyday experience as suggested by Popper — it is because we have a simple view that works for all everyday situations. Although this is dogmatism in the intellectual sense, it is not a true dogmatism in the attitude sense, because it is more like an inability, to conceive how an alternative view could be true. The alternative view was forced upon science by apparent empirical contradictions — how can the speed of light be constant (regardless of the velocity of the source or the receiver) without an ether? As seen from the following citation from part I of section IV of his *Enquiry Concerning Human Understanding,* Newtonian dogmatism is the sort of 'dogmatism' that Hume used to argue his scepticism:

> ALL the objects of human reason or enquiry may naturally be divided into two kinds, to wit, *Relations of Ideas*, and *Matters of Fact.* Of the first kind are the sciences of Geometry, Algebra, and Arithmetic; and in short, every affirmation which is either intuitively or demonstratively certain. *That the square of the hypothenuse is equal to the square of the two sides*, is a proposition which expresses a relation between these figures. *That three times five is equal to half of thirty,* expresses a relation between these numbers. Propositions of this kind are discoverable by the mere operations of thought, without dependence on what is anywhere existent in the universe. Though there never were a circle or triangle in nature, the truths demonstrated by Euclid would for ever retain their certainty and evidence.
>
> Matters of fact, which are the second objects of human reason, are not ascertained in the samemanner; nor is our evidence of their truth, however great, of a like nature with the foregoing. The contrary of every matter of fact is

> still possible: because it can never imply a contradiction, and is conceived by the mind with the same facility and distinctness, as if ever so conformable to reality. *That the sun will not rise to-morrow* is no less intelligible a proposition, and implies no more a contradiction than the affirmation, *that it will rise*. We should in vain, therefore, attempt to demonstrate its falsehood. Were it demonstratively false, it would imply a contradiction, and could never be distinctly conceived by the mind...

Hume used an example from Euclidian geometry as a demonstration of certain reasoning because it was impossible to imagine how Euclidian geometry could not be true. Einstein's theory of relativity questions whether Euclidean geometry is the true geometry. The idea that 'the speed of light is constant' is something like a logical contradiction. As we have seen, both Popper's and Hume's arguments against induction start out by effectively stating that we cannot get logical contradictions from empirical knowledge. And while there is no doubt that the only way we can get mathematically certain empirical knowledge is by mistake, there is also no doubt that we often do get extremely reliable science.

Should we reject a perpetual motion machine claim on the basis of the laws of thermodynamics? We should, unless we have got lots of spare time on our hands because we can achieve extremely high reliability with empirical laws. We should be dogmatic because dogmatic thinking can be efficient and reliable. However, we need to avoid a dogmatic *attitude*. If we are going to use the laws of thermodynamics to condemn a perpetual motion machine claim then we should make it clear that our confidence comes from a prejudice rather than a detailed look at the supposed perpetual motion machine.

3.2 Testability and Demarcation

Popper's demarcation problem is relevant to creationism because it is about the demarcation of science and non-science. Confirmation with respect to the demarcation problem is different to confirmation with respect to the problem of induction. Confirmation with respect to the demarcation problem is more about delusion arising from explanatory theories. There are three issues that allow the misuse of explanatory theories:

1. Many possible outcomes/predictions for one set of initial conditions.
2. The retrospective element associated with confirmation.
3. The attitude of confirmation in the sense of an objective.

The problem of induction asks why we feel justified in assuming the simplest description or why we should judge a pattern as being reliable because many other theories could explain the pattern. The problem of induction is one of having many explanations for a continuous pattern and trying to justify one particular explanation. While it is true that the word 'confirmation' often does imply checking what the theory says *after* we know the outcome of the experiment rather than predicting the outcome from the theory and then carrying out the experiment, confirmation in itself is not necessarily a problem. The observation of one non-white swan will falsify the law that 'all swans are white', providing that the definition of a 'white swan' is unambiguous.

The problem of explanations is the problem of having one explanation that says many things for one set of initial conditions. All predictive theories have explanatory power but an explanatory theory does not necessarily have predictive power because explanatory theories make probable predictions. Probable predictions invite an attitude of confirmation. This problem is essentially a confusion of the power to explain (or to make probable retrospective predictions) with the power to *predict*. What Popper is addressing is some sort of preferential thinking: Given certain conditions, we might say Freud's theory explains the reported behaviour A. Then we discover that the reported behaviour was wrong — the actual behaviour was B which is quite different from A. But then we argue that this other outcome also perfectly fits and confirms the theory. It may be that the theory does predict A and B but the question is whether we have lost sight of the fact that the theory predicts many things and whether we have been over-impressed by the confirmation.

The confusion Popper writes about occurs because we are not mindful of the number of other outcomes that will also fit the theory. And we are not mindful of this deficiency because we have an attitude of confirmation. Popper's solution to the *attitude* of confirmation is to test the theory.

We can find confirmations either because we are looking in the ad hoc region or because explanatory theories allow multiple outcomes. In either case, we can ignore or explain away inconsistencies and find consistencies in order to confirm a theory. The uncritical use of explanatory evidence allows the defence of a theory to be the direct objective whereas testing a theory helps to protect us from such an attitude. These two issues are outlined in Poppers first and second point in his *summary points*, given on p47&48 of *Conjectures and Refutations:*

- It is easy to obtain confirmations, or verifications, for nearly every theory, if we look for confirmations.
- Confirmations should only count if they are the result of risky predictions.

According to Popper, 'Every good scientific theory is a prohibition: it forbids certain things to happen'. It is easy to see how this requirement directly addresses explanatory theories that allow too many outcomes for the same initial conditions. We do not want a theory that predicts A,B,C for a given set of conditions but we want a theory that predicts A and prohibits every other alternative. In other words, we want a theory to be falsifiable, or refutable or testable. A theory that predicts all possible outcomes has nothing to say whereas a theory that predicts (a single outcome) often prohibits an infinite number of alternative possibilities.

So we test a theory and see if it passes; if it passes, we give it a tick. If it fails, we give it a cross. But what if we can't test the theory, and we can't even give it a tick or a cross? According to Popper, we should label it as 'unscientific'. Thus, for Popper, testability provides the demarcation between science and non-science. Given the fact that in science only observation and experiment may decide upon the acceptance or rejection of scientific theories (Popper calls this the *principle of empiricism*), there does seem to be some hope of saying (at least for some types of theories) that we can never have any evidence of the truth of the theory if the theory cannot be tested.

Although Popper's understanding says that there can be something very pseudo-scientific if not delusional in the *attitude* of confirming explanatory theories, it is important to note that there is nothing inherently wrong with explanatory theories since even a predictive theory will be an explanatory theory when we do not know all the initial conditions. The issue is that we can find confirmations to explanatory theories. Testing avoids the *attitude* of confirming a theory.

Regarding *theories*, Popper's intention was to only provide a 'not-testable in practice' categorisation. Popper claimed to be looking for a convention rather than a 'not-testable in principle' category. He claimed that he was looking for subjective value judgments rather than permanent classifications. Our theory might be about some condition in the centre of a star, but we do not currently have the instruments necessary to test the theory and therefore we should say that the theory is compatible with everything we *can currently* observe.

Popper complained about the fact that positivists used word categorisations against metaphysics. These positivists were people who attempted to develop the theory of knowledge called *logical positivism.* On page 12 of *The Logic of Scientific Discovery* Popper writes:

> Positivists usually interpret the problem of demarcation in a *naturalistic* way; they interpret it as if it were a problem of natural science. Instead of taking it as their task to propose a suitable convention, they believe they have to discover a difference, existing in the nature of things, as it were, between empirical science on the one hand and metaphysics on the other. They are constantly trying to prove that metaphysics by its very nature is nothing but nonsensical twaddle — 'sophistry and illusion', as Hume says, which we should 'commit to the flames'.
>
> If by the words 'nonsensical' or 'meaningless' we wish to express no more, by definition, than 'not belonging to empirical science', then the characterization of metaphysics as meaningless nonsense would be trivial; for metaphysics has usually been defined as non-empirical. But of course, the positivists believe they can say much more about metaphysics than that some of its statements are non-empirical. The words 'meaningless' or 'nonsensical' convey, and are meant to convey, a derogatory evaluation; and there is no doubt that what positivists really want to achieve is not so much a successful demarcation as the final overthrow and the annihilation of metaphysics.

Derogatory classifications are permanent classifications. Perhaps Popper's positivists had an attitude problem towards metaphysics.

Popper did not intend his philosophy to be used to provide absolute categorisations. For sure there are theories such as the theory that there are 'undetectable fairies in the garden' and the theory that 'all swans are white or not white' which are not science in an absolute sense, but we do not need Popper's philosophy to understand that there is something wrong with such theories.

I am not sure if Popper was consistent in his stance against the use of the not-testable categorisation to provide a permanent classification of a theory. Perhaps it would have been better if he said 'not yet testable and not yet science'. Another complication is that in the real world 'absolute categorisations' are not always absolute. In the real world it makes sense to use an 'is not' assertion even though there always will be some doubt about the best empirical assertions. Most of us would be happy to say that the theory that 'there are fairies in the garden' is not science in a meaningful sense because as explained in the notes we sometimes use certainty to mean sufficiently unlikely.

PART B:
Science and Creationism

Chapter 4 MISUSING POPPER

Part B is about the transition of western culture from the scientific revolution to killing all belief in a real God who can do things. On page 523 of his *History of Western Philosophy*, Russell explains the God-is-dead enlightenment:

> The solar system, at any rate, was kept going by its own momentum and its own laws; no outside interference was needed. There might still seem to be need of God to set the mechanism working; the planets, according to Newton, were originally hurled by the hand of God. But when He had done this, and decreed the law of gravitation, everything went on by itself without further need of divine intervention. When Laplace suggested that the same forces which are now operative might have caused the planets to grow out of the sun, God's share in the course of nature was pushed still further back. He might remain as Creator, but even that was doubtful, since it was not clear that the world had a beginning in time. Although most of the men of science were models of piety, the outlook suggested by their work was disturbing to orthodoxy, and the theologians were quite justified in feeling uneasy.

Russell sees the God-is-dead enlightenment as a natural extension of the scientific revolution. As discussed in another chapter, Russell commends ancient Greek atomism and the ancient Greek theory of evolution as being 'more scientific' than theories that invoked 'purpose in design' to explain.

Immanuel Kant (1724 – 1804) proposed the Nebular Hypothesis to explain the birth of the planets in 1755, when Laplace was only about 6 years old. This means there was a delay of almost 70 years after Newton's *Principia* before science headed in the direction of enlightened science. According to Anthony J. Abruzzo's article *The Origins of the Nebular Hypothesis* Descartes and other scientists proposed a Nebular Hypothesis before Newton's *Principia.* However part B is about how we used the scientific revolution to inspire enlightened thinking. Descartes' science and Descartes' rationalism resisted Newton's contribution to the scientific revolution.

From an ignorant perspective that sees only elliptical orbits and spherical bodies, there is room to allow that the planets could have been created as Kant and Laplace suggested. However, even from an ignorant perspective, it seems easy to appreciate that there was going to be some delay before science would insist that the creation of life should be explained by this 'more scientific' way of thinking about the world. Robert Chambers published his *Vestiges of the Natural History of Creation* in 1844. Alfred Wallace published a short essay on the theory of evolution in 1858 and Darwin published his *Origin of Species* in 1859.

On page 218 of their book *Darwin,* Adrian Desmond and James Moore explain how the scientific revolution inspired this new way of thinking about the world: "These discussions took place as Darwin was pondering the divine government himself. He too came to accept that 'the Creator creates by... laws.' Law ruled the earth, as it did the heavens; anything else was demeaning of God. Darwin complained that 'we can allow the satellites, planets, suns, universe, nay whole systems of universe[s,] to be governed by laws, but the smallest insect, we want to be created at once by a special act.' It was absurd; just as the winds in the Andes obeyed regular laws, so did the comings and goings of animals on the planet's surface. Dining at Leyell's, dancing at Babbage's, he found the idea of miraculous, catastrophic interruptions increasingly deplored.

The rule of law had to be upheld." Russell, Desmond, Moore and Darwin have the understanding that for science to be successful, science must kill the idea that there is a real intervening God. Desmond and Moore's description of who we were before the enlightenment (from pages 70-73) suggest that western science has been extremely successful:

> It is 1839. England is tumbling towards anarchy, with countrywide unrest and riots. The gutter presses are fizzing, fire-bombs flying. The shout on the streets is for revolution. Red evolutionists — visionaries who see life marching inexorably upward, powered from below — denounce the props of an old static society: priestly privilege, wage exploitation, and the workhouses. A million socialists are castigating marriage, capitalism, and the fat, corrupt Established Church. Radical Christians join them, hymn-singing Dissenters who condemn the 'fornicating' Church as a 'harlot,' in bed with the State.
>
> Even science must be purged: for the gutter atheists, material atoms are all that exist, and like the 'social atoms' people — they are self-organizing. Spirits and souls are a delusion, part of the gentry's cruel deceit to subjugate working people. The science of life — biology — lies ruined, prostituted, turned into a Creationist citadel by the clergy. Britain now stands teetering on the brink of collapse or so it seems to the gentry, who close ranks to protect their privileges...
>
> That same Saturday, 11 April, another moral lesson was being taught, this time to everyone in Cambridge: a young man was being led to his death on Castle Hill. Here, in the County Gaol, men accused of capital offences — some two hundred crimes carried the death penalty in 1829 — awaited trial in the Shire Hall, above the butchers' stalls in the market-place. The guilty had their sentences carried out on Castle Hill as an example to the community. Public

hanging was the ultimate deterrent in this world, just as hell-fire was in the next. The fear of the law and the fear of the Lord were all of a piece in Cambridge.

As Darwin shuddered at Fox's news, a noose was tightening. The execution had been well publicized. In an open letter, an evangelical clergyman had invited the whole county to attend and offer prayers 'for the unfortunate Culprit, that repentance and faith and so forgiveness may be granted him.' For 'highway robbery' a few miles west of town, William Osborne was hanged on the gallows...

On Thursday the 21st, the veteran radicals Richard Carlile and the Revd Robert Taylor tramped into Cambridge. They were notorious freethinkers, their reputations preceding them: proselytizers whipping up anti-Christian passions among the rabble. Carlile was a cobbler's son and a former tinplateman by trade. He had made his name as a fiery republican journalist and been given a six-year gaol sentence for blasphemous and seditious libel in the wake of the 1819 Peterloo Massacre (when innocent civilian demonstrators were cut down by cavalry). Taylor, his comrade-in-arms since his release, had himself just served a year for blasphemy. Blasphemy was a social crime because Christianity was part of the law of the land. By loosening the faith of 'illiterate' workers (as the Attorney General confirmed), Carlile and Taylor were lessening their ability 'to bear up against the pressure of misery and misfortune.' These demagogues, spokesmen for the secular working classes, were the bane of respectable society — clerical Cambridge not least.

On Tuesday, when Taylor and Carlile got wind of the vigilantes, they apologized to Smith, prodded the Vice-Chancellor again to restore his licence, and slipped out of town. Not wholly empty-handed: they had uncovered

> 'about fifty young collegians, who were somewhat bold in avowing Infidelity among each other' — even an Apostle or two perhaps. But only a handful of the students, the diehards knew, would ever 'break the shackles' of a Cambridge education and become out-and-out unbelievers. And they would have 'a most painful conflict to endure.' One such was Darwin. In later years he would remember Taylor as 'the Devil's Chaplain,' fearing that he himself might be similarly reviled, an outcast from respectable society, a terror to the innocent, an infidel in disguise. But for now he was safe — and ready to mend his ways.

Although spherical masses and elliptical orbits can be created naturally, contrary to Darwin's understanding, not just the smallest insect but the smallest self-sufficient life form seems like an extremely complex machine and complex machines seem as if they have been designed by a designer. Darwin was once persuaded by the design argument because there is something about the design argument that seems simple and obvious.

Fortunately, in Britain, a New Covenant enlightenment was getting off the ground at the same time that Darwin's God-is-dead enlightenment was getting off the ground. That Britain was full of so many very poor people, may have helped the New Covenant to prosper despite the message of science. It helped that Britain was not Roman Catholic. Whatever the explanation, not long after Newton, John Wesley (1703-1791) was transforming Britain with the New Covenant. The non-denominational London Missionary Society was formed in 1795. The British East India Company was ruling in India and fighting the opium wars with China and William Carey (1761-1834) and James Hudson Taylor (1832 –1905) were taking the Gospel to India and China. The British slave trade took slaves from Africa and David Livingstone took the Gospel to Africa. When Charles Darwin (1809 – 1882) was preaching that 'natural law did it', William Booth (1829 –1912) and Charles Spurgeon (1834 –1892) were preaching that God did it. When science was teaching British society that 'there is no evidence for a God who can do things because he has never done anything' Smith Wigglesworth (1859 – 1947) was raising the dead. Eventually though, it seems fair to say that science won the cultural war. Except for America perhaps, science has persuaded western society that, even if there is some kind of semi-real god who may have done some tinkering a few billion years ago, he is nothing like the God of the Bible.

Where Russell used the word 'scientific' and Desmond, Moore and Darwin used the term 'natural law', Professor Phillip Johnson used the term 'philosophical naturalism'. Philosophical naturalism is the prejudice and confidence that everything must happen by natural law. Without philosophical naturalism (or scientific thinking), Darwin would never have suggested that there was a natural explanation for life. Without philosophical naturalism, Russell, Darwin, and many others would never have considered that the God-is-dead enlightenment to be a natural extension of the scientific revolution.

Prejudice is a good word. On p32 of his *Philosophy of Biology,* Sober explains the real and ordinary credibility behind philosophical naturalism and why it is that even creationists believe that naturalism is probable. According to Sober, in order to explain the 'rumbling noises in the attic' we could suppose that there are 'Gremlins bowling in the attic'. Even though the Gremlin theory explains the noises, if scientists could come up with a 'scientific theory' that explains the noises, then scientists would not be very concerned whether the Gremlin theory is a good explanation. Scientists would not be very concerned about the Gremlin theory because Gremlin theories lack real and ordinary credibility.

Sober actually speaks in terms of the 'antecedent plausibility' of a theory rather than a 'scientific theory'. 'Antecedent plausibility' is about a 'prior probability' or a 'prejudice'. An event with a high prior probability is an event that we might predict even if our explanation did not need the event to happen. For the Gremlin theory to have a high prior probability, scientists would have to believe that Gremlins bowl in the attic even if there were no 'rumbling noises in the attic'. Science operates on the basis of many prejudices such as repeatability, the conservation of objects (10 bricks plus 10 bricks is equal to 20 bricks), the laws of thermodynamics and philosophical naturalism.

There is a religious form of philosophical naturalism, and a strong form of philosophical naturalism. In theory a religious scientists might be happy to say that the law of gravity could be understood as a supernatural or a natural law. The law of gravity can be understood as supernatural because as Hume argued we can never understand such things, and it can be understood as natural because it is predictable. In other words religious scientists have a problem with divine intervention. Religious scientists have a problem with a God like the God of the Bible because he does too many things.

There are many religious westerners who believe in a form of philosophical naturalism that allows that god did something very indirect in the distant past. This religious form of philosophical naturalism has some affinity to fuzzy 'religion factors' but not to the 'creation factor'. There are people who call themselves Christians who believe in some kind god even though they do not believe that God is required to explain creation because the 'creation factor' is not the only 'religion factor'.

When there was no explanation for life that we can all see around us then of course more people believed in a real God. However a fuzzy understanding of god has been around for a long time before Darwin. We can safely assume that some of the religionists in Desmond and Moore's account given above, had a fuzzy understanding of god. Aristotle believed in a supreme god who did not do anything. Buddhism ignored the 'creation factor' long before Darwin came up with his theory of evolution.

Richard Dawkins, adopts a strong form of philosophical naturalism, and therefore he has no choice but to understand that philosophical naturalism is a part of his thinking. Dawkins insists that science has a problem with all kinds of gods — even the do-nothing god of liberal Christianity. On page 15 of *The God Delusion*, Dawkins writes: "To expand the point, most of us happily disavow fairies, astrology and the Flying Spaghetti Monster, without first immersing ourselves in the books of Pastafarian theology, etc." Here, Dawkins is reflecting on the value of theology, but this is typical of Dawkins' approach to all things supernatural. Nevertheless, even Dawkins only believes in probable philosophical naturalism because he says that 'a creator almost certainly does not exist'.

It makes sense to use a definition of philosophical naturalism that puts creationists in a different category to people who believe in a religious form of philosophical naturalism because science is contending with creationists and most creationists are Biblical fundamentalists. It makes sense to speak in terms of the 'do nothing god' of science, because compared to the God of the Bible, the god of religious philosophical naturalism has done almost nothing. And given that, even creationists believe in probable naturalism, instead of defining philosophical naturalism as probable naturalism, it makes sense to define philosophical naturalism as a prejudice against the idea of 'a real God who can do real things'. With this definition we can say that the essential attitude of the God-is-dead enlightenment was philosophical naturalism.

Although part B is about the transition from the scientific revolution to the God-is-dead enlightenment, this chapter is not so much an argument against the God-is-dead enlightenment as it is an argument against arguments designed to end up at the right place. Discovery arguments are about progress whereas arguments that are designed to end up in the right place are about progress in a particular direction. Arguments that are designed to end up in the right place start from the answer and work backwards. On page 8 of their *Science and Creationism* booklet, the National Academy of Sciences complains about the fact that creationists arguments are designed to end up in the right place:

> The arguments of creationists are not driven by evidence that can be observed in the natural world. Special creation or supernatural intervention is not subjectable to meaningful tests, which require predicting plausible results and then checking these results through observation and experimentation. Indeed, claims of "special creation" reverse the scientific process. The explanation is seen as unalterable, and evidence is sought only to support a particular conclusion by whatever means possible.

Most creationists are Biblical fundamentalists and we fundamentalists always want our arguments to end up in the right place.

Although it is true that people like me spend their time thinking and arguing about why the Bible is true, this chapter is about the fact that, when it comes to the topic of creationism, there are many individuals who regard themselves as scientific thinkers, critical thinkers, freethinkers and enlightened thinkers who want their arguments to end up on philosophical naturalism. I like to refer to people who understand that their arguments must end up agreeing with philosophical naturalism as freethinkers. This is because it seems to me that freethinkers are people who are not allowed to think about the possibility of a real God.

For many scientists the 'creationism is not science' assertion ends up in the right place. For many scientists the fixed doctrines of the Bible cannot be science. Although it seems difficult to explain how the specifics of Biblical creationism could be so confusing, the Discovery Institute (originally including Phillip Johnson, Michael Behe and William Dembski) have achieved some success against the theory of evolution using *intelligent design theory.* Intelligent design theory focuses on the essentials of creationism and ignores all the specifics of Bible creationism. Intelligent design theory has had some influence on the American school curriculum.

Spokesmen for evolution apparently frustrated by the fact that intelligent design theory is not as easy to debunk as Biblical creationism, also like to say things like 'intelligent design theory is not this or not that'. And the fact that people sometimes say that 'intelligent design is not testable therefore it is not science', means that some people are basing their judgments on Popper's testability philosophy.

4.1 Pseudo-Science

Popper did not intend for his 'not testable' categorisation to be used as a 'not-testable in principle' categorisation against theories. Popper did not intend that a theory that is tested and absolutely falsified to be 'more scientific' than a theory that has not yet been tested because of a lack of adequate test instruments. However, Popper did want to address a *delusional attitude* associated with explanatory theories. On pages 45-46 of Conjectures and Refutations, Popper explains the *attitude* of 'misusing explanatory theories and finding evidence bias to confirm our thinking':

> I found that those of my friends who were admirers of Marx, Freud, and Adler, were impressed by a number of points common to the theories, and especially by their apparent explanatory power. These theories appeared to be able to explain practically everything that happened within the fields to which they referred. The study of any of them seemed to have the effect of an intellectual conversion or revelation, opening your eyes to a new truth from those not yet initiated. Once your eyes were thus opened you saw confirming instances everywhere: the worldwas full of verifications of the theory. Whatever happened always confirmed it. Thus its truth appeared manifest; and unbelievers were clearly people who did not want to see the manifest truth; who refused to see it, either because it was against their class interest, or because of their repressions which were still 'un-analysed' and crying out for treatment.
>
> The most characteristic element in this situation seemed to me the incessant stream of confirmations, of observations which 'verified' the theories in question; and this point

> was constantly emphasized by their adherents. A Marxist could not open a newspaper without finding on every page confirming evidence for his interpretation of history; not only in the news, but in the presentation — which revealed the class bias of the paper — and especially of course in what the paper did not say. The Freudian analysts emphasised that their theories were constantly verified by their 'clinical observations'. As for Adler, I was much impressed by a personal experience. Once, in 1919, I reported to him a case which to me did not seem particularly Adlerian, but which he found no difficulty in analysing in terms of his theory of inferiority feelings, although he had not even seen the child. Slightly shocked, I asked him how he could be so sure. 'Because of my thousandfold experience' he replied; whereupon I could not help saying: 'And with this new case, I suppose, your experience has become thousand-and-one-fold.'
>
> What I had in mind was that the previous observations may not be much sounder than this new one; that each in its turn had been interpreted in the light of 'previous experience', and at the same time counted as additional confirmation. What, I asked myself, did it confirm? No more than that a case could be interpreted in the light of the theory. But this meant very little, I reflected, since every conceivable case could be interpreted in the light of Adler's theory, or equally of Freud's. I may illustrate this with two very different examples of human behaviour: that of a man who pushes a child into the water with the intention of drowning it; and that of a man who sacrifices his life in an attempt to save a child. Each of these cases can be explained with equal ease in Freudian and in Adlerian terms.

The theory of evolution is an explanatory theory rather than a predictive theory. When we say that some theory explains some facts we mean that given the theory we should not be surprised by the facts. An explanation is a probable prediction made after the event. We can say that a theory provides an explanation for an outcome if the outcome is compatible with the theory, but there could be other outcomes that are compatible with the theory and so we need to ask us how probable was the agreement with the theory. On page 20 of his book *Darwin's Dangerous Idea — Evolution and the Meanings of Life* Dennett writes:

> The fundamental core of contemporary Darwinism, the theory of DNA-based reproduction and evolution, is now beyond dispute among scientists. It demonstrates its power every day, contributing crucially to the explanation of planet-sized facts of geology and meteorology, through middle-sized facts of ecology and agronomy, down to the latest microscopic facts of genetic engineering. It unifies all of biology and the history of our planet into a single grand story. Like Gulliver tied down in Lilliput, it is unbudgeable, not because of some one or two huge chains of argument that might—hope against hope—have weak links in them, but because it is securely tied by hundreds of thousands of threads of evidence anchoring it to virtually every other area of human knowledge. New discoveries may conceivably lead to dramatic, even "revolutionary" shifts in the Darwinian theory, but the hope that it will be "refuted" by some shattering breakthrough is about as reasonable as the hope that we will return to a geocentric vision and discard Copernicus.

It could be that Dennett's unbudgeable confidence in philosophical naturalism causes him to *find* many threads of evidences and it causes him to ignore the probable nature of the evidence as explained by Popper. Had Dennett given sufficient details to allow an examination of his explanatory justifications, Popper would advise that we should question whether we would have found something wrong with the theory of evolution had the 'thousands of threads of evidence' been different. Because we do not have the details, and assuming that Dennett does not say things like 'creationism is not testable therefore it is not science', all we can say for sure is that Dennett does not acknowledge the delusional power of explanatory thinking and that he is ignoring Popper.

The fundamental reason why explanatory arguments have this delusional power is that they 'predict' more than one outcome. Explanatory arguments are probable arguments. The probable nature of an explanatory argument can come directly from the incompleteness of the theory. Theories of psychology might always be explanatory theories simply because they will never be able to account for all the relevant differences in individuals, but we may be able to use such theories to make probable predictions or predictions about population averages. Alternatively the probable nature of an explanation could be the result of a probable auxiliary assumption that relates to some initial conditions — in one setting a theory might make a singular prediction but in another setting the theory might allow many different outcomes because some of the initial conditions are unknown. In this case we could adopt probable auxiliary assumptions that specify all the relevant initial conditions so that we can use the theory to provide probable predictions.

A probable conjecture is an explanation except that when we are conjecturing, we determine the auxiliary assumptions before we see the results. If the number of likely auxiliary assumptions were limited to a small number then there would be a chance of a probable prediction based on some assumptions.

Unlike Dennett, *NAS (National Academy of Sciences* of America) specify some of the details of their explanatory arguments. On page 17 of their booklet *Science and Creationism NAS* writes:

> Similarly, a wide variety of organisms from fruit flies to worms to mice to humans have very similar sequences of genes that are active early in development. These genes influence body segmentation or orientation in all these diverse groups. The presence of such similar genes doing similar things across such a wide range of organisms is best explained by their having been present in a very early common ancestor of all of these groups.

It the above, NAS uses the phrase 'best explained'. Best explanation theories were not the focus of Popper's testability criterion. Best explanation theories can provide extreme reliability when we have a handle on all possible explanations. However NAS needs to say why this seems like an unlikely outcome given creationism.

While it is true that evolution can be used to explain such things, there could be auxiliary assumptions hidden somewhere within the argument, because the theory of evolution is not a predictive theory. Each set of auxiliary assumptions results in a different version of the theory of evolution. According to Popper's thinking, *NAS* should be asking themselves what they would conclude if the did not find such things. *NAS* should be asking themselves what would they expect to find if the theory of evolution is the truth and what should they expect to find if creationism is the truth. On the same page *NAS* also argue that:

> The code used to translate nucleotide sequences into amino acid sequences is essentially the same in all organisms. Moreover, proteins in all organisms are invariably composed of the same 20 amino acids. This unity of composition and function is a powerful argument

in favor of the common descent of the most diverse organisms.

In order for NAS to suggest that this is a powerful argument that *NAS* must have in mind some robust auxiliary assumptions. The data is only predicted by the version of evolution that says:

1. All life came from one ancestor — life did not originate from two or more independent sources (This seems to agree with the understanding that the creation of life is a difficult problem).

2. The code was passed on from one life form to another without any changes whatsoever from the first life form to all other life forms. This requires the complete code to be perfected in bacteria before the bacteria evolved into something else. And it requires that all bacterial strains with a different code became extinct.

NAS have said that the outcome is a powerful argument in favour of common descent but in fact it is only predicted by a very specific version of the theory of evolution. It seems that this version was introduced after the data was noted and for all we know may change with the discovery of new data.

There are problems with these auxiliary assumptions. As seen in the table below, the DNA code is ordered with the second and first positions in the codon being the most important positions for dictating which amino acid (or stop or start instruction) is signalled for. A change in the second and the first positions will nearly always result in a change in the amino acid signalled by the codon. Though a change in the second position of the codon for the stop signal and a change in the first position for the codons for Arginine and Leucine will not necessarily produce a change, in all other cases a change in these positions will produce a change. The situation for the third position for the triplet codon is quite different because except for the start codon, one version of the stop codon and the codon for Tryptophan, a change in the third position may mean no change in the amino acid produced. And in 8 of the first and second letter combinations a change in the third position cannot produce any change.

TABLE OF THE GENITIC CODE

First Position	Second Position				Third Position
	U	C	A	G	
U	Phenylalanine	Serine	Tyrosine	Cysteine	U
	Phenylalanine	Serine	Tyrosine	Cysteine	C
	Leucine	Serine	Stop	Stop	A
	Leucine	Serine	Stop	Tryptophan	G
C	Leucine	Proline	Histidine	Arginine	U
	Leucine	Proline	Histidine	Arginine	C
	Leucine	Proline	Glutamine	Arginine	A
	Leucine	Proline	Glutamine	Arginine	G
A	Isoleucine	Threonine	Asparagine	Serine	U
	Isoleucine	Threonine	Asparagine	Serine	C
	Isoleucine	Threonine	Lysine	Arginine	A
	Methionine/ Start	Threonine	Lysine	Arginine	G
G	Valine	Alanine	Aspartic acid	Glycine	U
	Valine	Alanine	Aspartic acid	Glycine	C
	Valine	Alanine	Glutamic acid	Glycine	A
	Valine	Alanine	Glutamic acid	Glycine	G

The order in the code requires an explanation. If the arrangement is the best arrangement, then this means that either a designer selected the best arrangement or evolution selected the best arrangement but it also means that *NAS* have a problem with the common ancestor explanation to explain the uniformity of the code, since if there is a best arrangement then even if there were multiple original ancestors, we might expect them all to have the same best arrangement. The conjecture about the 'common ancestor' is itself an auxiliary assumption, which is in turn attached to another auxiliary assumption namely that 'there is no reason to have a particular arrangement of the code' or that 'there is no best arrangement of the code'. If we assume evolution and we assume there is a best arrangement, then we would have no basis for assuming a common ancestor. This means that given evolution, the 'best arrangement' and the 'common ancestor' assumptions are conflicting assumptions. My guess is that the 'common ancestor' idea was not a likely auxiliary assumption until after the code was found to be the same for all life forms because *NAS* are confirming evolution in the sense of attitude. This is to say that had *NAS* not known about the common code they may have considered it to be a possibility, but they would not have predicted it with any degree of confidence. It seems that *NAS* are working backwards from the primary prejudice of science.

A code is also used in the mitochondria of cells that is different from the code used by the chromosomes in the nucleus. And not only is the code different from the nucleus code, but different life forms do not necessarily have the same mitochondrial code. Science could explain this by saying that different codes are more suited to different life forms and these codes have been selected by evolution. Thus, if we see some evidence that fits our theory, then we mention it, but if we see some evidence that does not line up with our theory, we assume that it can be explained. The two factors we need for this kind of science are 'retrospective confirmation' and 'probable evidence' because these are the two factors we need for the *finding-evidence* bias.

If we want to avoid the finding-evidence bias we should determine the auxiliary assumption we are most comfortable with prior to testing the theory. Otherwise, we might select our auxiliary assumption to match the explanation with the known data without acknowledging our backwards approach. An explanation can look like a confirmation if we fail to identify the probable nature of the auxiliary assumptions that are attached to the explanatory arguments. If there were two codes, what would *NAS* say about the theory of evolution? What if there were one hundred codes? What would *NAS* say if we found that there were slight differences in the codes of various species?

Suppose we observe a simple pattern in some data and we would like to explain the pattern. When we say some explanatory argument explains the pattern, then we are saying that there is no inconsistency between the theory and the pattern providing we choose a certain auxiliary assumption. If we judge that the auxiliary assumption is probable, then we might be able to say that the pattern seems reasonably likely given the theory — the explanation and the auxiliary assumption answers the question. However in assessing the theory we must remember that the strength of the explanatory argument will only be as good as the probable truth of the auxiliary assumption. And to avoid a circular argument, the probability value we place on the auxiliary assumption must be the value prior to the explanation. In other words the probability value we place on the auxiliary assumption must be the value we determine assuming that creationism is science.

Regardless of the reliability of our scientific prejudices, we can only use our prejudices to openly make assertions against creationism. If for example, *NAS* were commissioned to investigate whether or not there are Gremlins in the attic, *NAS* would somehow need to divorce itself from the confidence that there are no Gremlins in the attic, otherwise why bother to investigate. The reason such an attitude might not be such an easy thing to adopt is because we are all extremely confident that there are no Gremlins in the attic.

The reason why *NAS* judges the auxiliary assumptions as probable is because they do not keep it independent of the theory, but they use the probability of the theory to determine the probability of the auxiliary assumption. The fact that philosophical naturalism is a foregone conclusion makes it very easy for all of us to use the probability of the 'auxiliary assumption given naturalism' for the probability of the 'auxiliary assumption'.

In the above citation, we see that NAS used the phrase 'a powerful argument in favor of' without considering whether or not creationism can also explain the universal code. The reason we so easily go along with the idea that some evidence can establish evolution without considering whether or not the evidence is compatible with creationism, is that we all think that naturalism is so likely and that creationism is so unlikely. Why should we bother to investigate if there are any Gremlins in the attic when we know that there are no Gremlins in the attic?

Unless we test our theory then the assessment of the number of alternative predictions that are compatible with our theory and the assessment of the probability of the auxiliary assumption and the assessment of whether we have found the evidence may require some onerous critical thinking and some clear headed subjective judgments — and none of us are very good at these things. Not that there is anything wrong with confirmations except that they sometimes provide opportunity for attitude error because:

1. Probable means that there will always be some evidence that agrees with our theory and some evidence that disagrees with our theory.

2. Retrospective means that there is opportunity to confirm (in the sense of an objective) or rather to find evidence for the theory — assuming that either the theory makes probable predictions or that we have a probable auxiliary assumption.

3. Explanatory power can be misleading because an uncritical attitude allows us to use the consistency

> between the theory and the data to inappropriately confirm the theory. We can either find data that is consistent with the theory or we can interpret the data to be consistent with the theory. This seems to be the same as saying that we fail to recognise that the prediction has a very low probability or we misjudge the probability of the auxiliary assumption, as per the 'universal code' example discussed above. Although spokesmen for evolution like to use Popper's testability theory to say that 'creationism is not science' in the sense of a derogatory judgment that Popper did not intend, Popper did intend that we use his testability philosophy to escape the fact that it is easy to be over-impressed by the confirmation provided by explanatory arguments. On page 34 of his *Conjectures and Refutations*, Popper writes: "What, I asked myself, did it confirm? No more than that a case could be interpreted in the light of the theory. But this meant very little, I reflected, since every conceivable case could be interpreted in the light of Adler's theory, or equally of Freud's."

Freethinking confuses 'best explanation thinking given philosophical naturalism' with ordinary best explanation thinking because philosophical naturalism is not aware that philosophical naturalism is an assumption. It is difficult to investigate if there are Gremlins in the attic when we all know that there are no Gremlins in the attic and it is difficult to investigate if creationism is science if we know that creationism is not science.

4.2 Creationism is Not Science?

Spokesmen for evolution say things like 'creationism is religion not science' apparently because they think that by putting creationism into the religious category we can learn something about the credibility of creationism. Spokesmen for evolution seem to be happy to say that creationism is not science because it has religious motivations or because it is a fixed doctrine of the Bible. Spokesmen for evolution seem to be convinced that they do not need to consider the credibility of intelligent design because like Descartes they think they can use words-and-definitions to decide empirical truth.

Newton's theories were threatened by the assertion that 'theories that conjectured magical forces at a distance rather than force by contact explanations are not science'. According to page 25 of the NAS *Science and Creationism* booklet:

> The claim that equity demands balanced treatment of evolutionary theory and special creation in science classrooms reflects a misunderstanding of what science is and how it is conducted. Scientific investigators seek to understand natural phenomena by observation and experimentation. Scientific interpretations of facts and explanations that account for them therefore must be testable by observation and experimentation. Scientific investigators seek to understand natural phenomena by observation and experimentation. Scientific interpretations of facts and the explanations that account for them therefore must be testable by observation.
>
> Creationism, intelligent design, and other claims of supernatural intervention in the origin of life or of species are not science because they are not testable

> by the methods of science. These claims subordinate observed data to statements based on authority, revelation or religious belief. Documentation offered in support of these claims is typically limited to special publications of their advocates. These publications do not offer hypothesis subject to change in light of new data, new interpretations, or demonstration of error. This contrasts with science, where any hypothesis or theory always remains subject to the possibility of rejection or modification in the light of new knowledge.
>
> No body of beliefs that has its origin in doctrinal material rather than scientific observation, interpretation, and experimentation should be admissible as science in any science course. Incorporating the teaching of such doctrines into a science curriculum compromises the objectives of public education.

The first paragraph states what we want from science. The second and third paragraphs attempt to state why creationism is not science. According to NAS, creationism is not testable by the methods of science.

The *Science and Creationism* booklet does not say why creationism is not testable so the reader would assume the reason to be obvious. Also it seems that their claim is not just that creationism is not-testable but that it is not-testable in principle. A not-testable in principle claim asserts that there is something in the nature of the theory (rather than the inadequacy of the test instruments) that makes it not-testable.

One obvious reason that creationism might not be testable in principle and therefore is not science is because creationism includes a supernatural component. This seems like a reasonable interpretation because the word ‘supernatural’ is such an unusual word in a scientific context that it easily associates itself with ‘not science’.

Perhaps because the empirical factor (observation and experimentation) is first mentioned, we might say that because the scope of science is empirical, creationism does not fall within the scope of science. Creationism involves the supernatural, therefore, it is not within the scope of science, because science can only test and understand natural things.

However, when people use a scope criterion or some defining property of science to say what is science, they are saying that creationism is not science in the same sense we might say that mathematics is not science. Suppose that God did create all life and that there is evidence of design in the genetic code. We could say (using the scope criterion) that such evidence of design is religion not science. Or, suppose that evolution is testable and that it is tested and falsified — then by elimination of evolution, then creationism would be the logical explanation. We would then say 'creationism is not science but it is the truth' and 'evolution is science but it is not the truth'. There are perhaps too many possibilities to consider but the one thing that we know for sure is that *NAS* do not want the 'not science' judgement to speak against the credibility of evolutionary science.

We cannot solve these sorts of problem with a definition because a definition puts something else before credibility. From p30 of *Science and Creationism*:

> Scientists as well as educators have concluded that evolution — and only evolution — should be taught in science classes because it is the only *scientific* explanation for why the universe is the way it is today.

Although it is true that evolution is the only scientific explanation, there is a difference between teaching 'evolution is the only scientific explanation' and teaching our children that evolution is credible. The intention of the *NAS* is to teach school children that 'evolution is reliable truth' rather than 'evolution is the only scientific explanation'. And again to insist on a scientific explanation (or some other definition) is to put credibility in second place. The only reason we can ever have for teaching children that evolution is reliable science is if the arguments favour evolution and not because evolution is the only scientific explanation. A scientific response to the claim that Jesus rose from the dead is that 'he was not really dead'.

Although there is a sense in which we can say that creationism is not science, if we want to avoid the problem of saying that 'creationism is not science but it is true', then credibility must be what we use to determine what is science and what is not science.

If we could say that there is no evidence for creationism (say because creationism does not have anything to say), then we will never be able to test creationism to provide any other evidence for creationism, then there would be some possibility of linking testability with credibility. If, for some reason we could never test a theory against reality, then it might seem that we cannot have any evidence that a theory is true. However that sort of thinking does not apply to theories that are based on other theories. Theories that are used to provide an explanation within the framework of existing empirical knowledge, can be probable without testing. Creationism and evolution are more like theories that explain why we get dew on the grass or why do we only see one side of the moon, and such theories can be credible without the help of testing.

It could be that the *NAS* 'creationism is not science' statement comes from the fact that creationism is a supernatural theory. However, there are two aspects to creationism — one is the how of creation (which science will never understand) and the other is the fact of creationism — whether or not it happened. It should be that *NAS* understands this to be the case because on page 28, they argue something similar for evolution:

> The theory of evolution explains how life on Earth has changed. In scientific terms, "theory" does not mean "guess" or "hunch" as it does in everyday usage. Scientific theories are explanations of natural phenomena built up logically from testable observations and hypotheses. Biological evolution is the best scientific explanation we have for the enormous range of observations about the living world.
>
> Scientists most often use the word "fact" to describe an observation. But scientists can also use fact to mean something that has been tested or observed so many times that there is no longer a compelling reason to keep testing or looking for examples. The occurrence of evolution in this sense is a fact. Scientists no longer question whether descent with modification occurred because the evidence supporting the idea is so strong.

The how of supernatural creation is outside the scope of science. The creator and his supernatural workings may be outside the scope of science, but he left creation behind for us to detect. Science may never understand the process of creation by God and therefore there is a sense in which we can say that creationism is not science. The real issue is whether or not it is reasonable to invoke some sort of creator solution to explain creation.

People also say that intelligent design has nothing to say. It is true that intelligent design has less to say because intelligent design says 'creation was the work of an intelligent designer but who knows who the designer was'.

However intelligent design theory must have something important to say otherwise it would not be able to upset so many spokesmen for evolution. Intelligent design theory recognises that naturalism does not have power to create complexity. Intelligent design recognises that only a designer can design complexity. Because Fred Hoyle and Francis Crick suggested that aliens were somehow responsible for getting life going on earth, it makes sense to identify Fred Hoyle and Francis Crick as intelligent design theorists. Hoyle and Crick were unusual because in spite of their commitment to the primary prejudice of science, they were able to recognise that 'the evidence for a creator is manifest in creation'.

People also say that intelligent design is religion not science. Perhaps these religion-not-science statements are popular because government bodies must be religiously neutral. Science is different to religion in scope, but when people say something is religion not science then they are assuming that it is not possible for religion to overlap science. The simple and obvious fact is that theories of creation are both science and religion. It is also true that for science we want scientific reliability whereas for religion at best we have a different sort of reliability. We can say that some specific version of creationism, say the idea that God created life *in 6 days* is not science because the idea that 'God created all life forms *in six days*' will never have independent scientific credibility, because it is far too specific. It may be that the specifics of Biblical creationism help spokesmen for evolution to believe that there is sense in saying things like 'creationism is not science because creationism is not testable'.

4.3 Creationism is not Testable?

Even though testability does not work as a criterion of demarcation, it helps to note that there have been spokesmen for evolution who have claimed that 'evolution can explain either imperfection or perfection but creationism is only consistent with or only predicts perfection'. For example according to page 81-82 of Christopher Hitchens' *God is not Great*, Dr Shermer claims that: "The anatomy of the human eye, in fact, shows anything but "intelligence" in its design. It is built upside down and back wards, requiring photons of light to travel through the cornea, lens, aquaeous fluid, blood vessels, ganglion cells, amacrine cells, horizontal cells, and bipolar cells before they reach the light sensitive rods and cones that transduce the light signal into neutral impulses-which are then sent to the visual cortex at the back of the brain for processing into meaningful patterns. For optimal vision, why would an intelligent designer have built an eye upside down and backwards?" Shermer thinks that creationism is testable and that he has falsified creationism. Similar thinking once concluded that both the tonsils and appendix in people are redundant remnants of evolution. On page 52 of *Philosophy of Biology* Elliot Sober argues that 'creationism is not testable':

> The hypothesis that God separately created living things is testable only when it is conjoined with auxiliary assumptions. But how is one to know which auxiliary assumptions to believe? Paley saw God one way; other creationists may prefer a different picture of what God would be like. Different religions conceive of God in different ways. And there are conceptions of God (like the trickster God discussed before) that perhaps are not

> part of any mainstream religion. How is one to choose? The fact that some of these conceptions of God are familiar while others are decidedly odd is no basis for selecting. What one wants is evidence that one of them is true and the rest are false. Without any evidence of this sort, the project of testing the hypothesis that God separately created the species that populate the living world is stopped dead.

According to Sober then, the reason why creationism is not testable is because it is too general to say anything — or rather it is so general it can be made to allow everything. The fundamental requirement of Popper's testability criterion is that a testable theory must say something that can be tested. Creationism is just a word — depending on who is using it, it might mean instantaneous creation or divinely assisted evolution. And the fact that Sober's testability has nothing to do with inadequacy of the test instruments suggests that there is an absolute categorisation looming.

There is, in fact, some common sense understanding in Sober's thinking because there are numerous auxiliary assumptions that can be attached to creationism. People forget that according to Biblical creationism God modified his creation when sin entered the world. In the original creation, there were no carnivores (and no disease). According to Romans 8:22 from the time that sin entered the world the entire world has been under a curse: "For we know that the whole creation groaneth and travaileth in pain together until now." God apparently modified the herbivores to provide a defence against carnivores, and apparently he modified plant life as well. The snake of the garden of Eden walked on legs and spoke.

When I argue the specifics of Scripture, rather than sticking to the generalities of intelligent design theory then this raises the question of the influence of cultural prejudices on our judgements. My creationism includes Sober's creationism that he describes as the 'hypothesis that God separately created the species that populate the living world', but it does not include the following versions of creationism (from pp 7 & 24 of *Science and Creationism*):

> Many religious persons, including many scientists, hold that God created the universe and the various processes driving physical and biological evolution and that these processes then resulted in the creation of galaxies, our solar system, and life on Earth. This belief, which sometimes is termed "theistic evolution," is not in disagreement with scientific explanations of evolution. Indeed, it reflects the remarkable and inspiring character of the physical universe revealed by cosmology, paleontology, molecular biology, and many other scientific disciplines...
>
> Opinion polls show that many people believe that divine intervention actively guided the evolution of human beings. Science cannot comment on the role that supernatural forces mightplay in human affairs. But scientific investigations have concluded that the same forces responsible for the evolution of all other life forms on Earth can account for the evolution of human beings.

In the above, *NAS* seem ambivalent towards divinely guided evolution and even somewhat friendly towards theistic evolution. This is because science only has a problem with a real God who can do things.

Theistic evolution has a god who created the laws so that the universe could evolve essentially by itself with perhaps a little subsequent tinkering here and there if required. The divinely guided evolution theory is similar except that it has more emphasis on the tinkering stage. Both theories accept the power of the mutation-selection-copy process to create complexity and some versions accept that naturalism created the first life form.

Although there is some common sense understanding in Sober's thinking, the same common sense says that there are versions of creationism which we naturally regard as more probable, because credibility is about prior probabilities or what Sober calls 'antecedent plausibility'. On page 316 of *The Blind Watchmaker*, Dawkins writes: "Nearly all peoples have developed their own creation myth and the Genesis story is just the one that happened to have been adopted by one particular tribe of Middle Eastern herders. It has no more special status than the belief of a particular West African tribe that the world was created from the excrement of ants. All these myths have in common that they depend upon the deliberate intentions of some kind of supernatural being." It could be difficult to get enlightened science to *say* the Bible version of creation has more credibility than the idea that the world was created from the excrement of ants. Fortunately we can avoid that issue, because a common sense approach favours testing the more troublesome version of creationism.

Biblical fundamentalism is more troublesome because it denies the primary prejudice of science. Biblical creationists argue that creation is evidence for a real creator doing real things. Biblical creationists deny that evolution or anything like evolution happened or could happen. And only Biblical creationist have Biblical reasons for why the evolution solution should be debunked and why creationism should be taught in the schools. A practical approach says that what matters the most to science is the elimination of the predominant forms of creationism that spring from Biblical fundamentalism.

Although Popper's 'not-testable therefore not-science concept' adds to the confusion, testing a theory rather than confirming a theory is a good way of evaluating the evidence. There are, however, various forms of Biblical creationism.

Some Biblical creationists believe in a young earth and some allow that the earth is very old. However, nearly all creationists believe that the life we see on earth is only about 6000 years old. The 'young life' assumption seems like a reasonable auxiliary assumption for creationists. Such an auxiliary assumption will catch most biblical creationists. There is then the matter of how we test for this 'young life' assumption.

Once the more troublesome versions of creation are eliminated then, as Sober says, creationism has not been falsified because it will always be possible to construct a new type of creationism that fits the evidence, but the new version of creationism might not present a challenge to the theory of evolution. We could suppose that we have a trickster God who wanted things to look like they had evolved, but most of us would be content to judge that sort of creationism as unlikely.

All molecules (such as various proteins) directly constructed from the DNA code are a sequence of amino acids. There are only twenty standard amino acids. However, the DNA code itself is a sequence of one of four molecules that can be identified by the letters U C A and G and a sequence of three of these letters will code for one amino acid. Because there are only twenty amino acids (and a start and stop signal) and sixty-four different codons that code for these molecules, there are a lot of neutral mutations possible in the DNA. If all mutations are equally likely, then approximately two-thirds of all mutations should be neutral. A portion of these neutral mutations will accumulate because they do not have a detrimental effect and because of the different time scales there should be vastly different amounts of neutral mutations predicted by the evolution model against the creation model. There is no easy escape for either theory, and this should be sufficient to challenge the 'fact' of evolution. Perhaps we could do similar tests with pseudo-genes, assuming what we understand of pseudo-genes does not come from our confidence in evolution.

Testing creationism and evolution is about determining whether creationism or evolution is the truth, whereas this chapter is about using words-and-definitions assertions, philosophical naturalism and the finding-evidence bias to end up at the right place.

Testing a theory is supposed to help us escape the need to end up at the right place. Testing is not supposed to be a way of confirming a theory. Although the tests suggested above should help us to answer some questions, I already suspect that such tests will support creationism because I believe that I read about these tests on someone's website many years ago. We undermine the purpose of testing when we find our tests, because testing is supposed to provide a way around the problem of using finding-evidence bias to ensure that our arguments end up in the right place. That does not mean the tests must be scrapped but it does mean that unless we are careful then we can use testing as another way of confirming what we know to be true. Such an attitude is not science.

The use of words and definitions like 'creationism is not science' in order to make creationism go away is not science. Even words like the 'fixed doctrines of the Bible' can be interpreted as a derogatory assertion by scientists who believe in the magic of words-and-definitions. When Christians preach the Gospel, people respond with arguments like 'religion is the cause of many wars' as if the fact that religion has been the cause of many wars is evidence that Jesus did not rise from the dead. I see the need to mention this problem because people who are fond of 'derogatory logic' are bound to be sensitive to 'derogatory logic'. People who believe in derogatory logic will have a problem distinguishing arguments that seem derogatory of science from arguments that undermine the truth of science.

If a fundamentalist like myself suggests that science is anti-God (where God is a real God like the God of the Bible) then scientists who believe in derogatory logic may see it as their duty to explain such a statement away rather than discovering any truth behind the statement. Regardless that the doctrine that 'the stars and planets are not supported by crystalline spheres' speaks against the idea of a God who can do things, if the stars are not supported by a crystalline spheres then science must report that the stars are not supported by a crystalline sphere. Regardless that the doctrine that 'there is no empirical evidence for a God who can do things' is the lesson of the theory of evolution, if there is no evidence for a God who can do things then science must report that 'there is no empirical evidence for a God who can do things'.

Providing we mean 'God' to be a God like the God of the Bible then not only is philosophical naturalism anti-God in a practical sense but it is anti-God in a spirit-and-motivation sense. Although scientific facts that speak against the idea of a real God who can do things are not necessarily anti-God in spirit-and-motivation, the doctrine that 'their is no God who can do things' is anti-God in spirit-and-motivation, and modern science is built on this understanding.

Philosophical naturalism is the *primary prejudice of science* when the topic is evolution or creationism. Science operates on the basis of many prejudices such as repeatability, the conservation of objects (10 bricks plus 10 bricks is equal to 20 bricks) and the laws of thermodynamics and although philosophical naturalism is no more important than these other prejudices, when the topic is evolution it makes sense to identify philosophical naturalism as the primary prejudice of science. For example on page 250 of *Darwin's Dangerous*

Idea, Dennett, who is a non-religious scientist, writes of one of Dawkins assumptions:

> Does Dawkins know this? Does he know that the postulated intermediates were less fit? Not because he has seen any data drawn from the fossil record. This is purely a theory-driven explanation, argued a priori from the assumption that natural selection tells us the true story some true story or other — about every curious feature of the biosphere. Is this objectionable? It does "beg the question" but what a question it begs! It assumes that Darwinism is basically on the right track. (Is it objectionable when meteorologists say, begging the question against supernatural forces, that there must be a purely physical explanation for the birth of hurricanes, even if many of the details so far elude them?)

There are little gaps and there are big gaps in our understanding. Dennett seems happy to use a hair-line fracture, to represent the Grand Canyon. The reason that Dennett is happy to do that is because they are both gaps, but the real point to this argument is that we can be sure that nobody would not reverse this analogy. That is, in order to justify the understanding that we should exclude supernatural forces from our understanding of hurricanes, nobody would use the gap of creation to explain away the gaps in our understanding of hurricanes. This example should mean that scientists who do not believe in derogatory logic understand that philosophical naturalism is the primary prejudice of science when the topic is evolution.

Of course it is in some sense natural that science is only concerned about faulty arguments that end up at the wrong place. Even though the God of the Bible would disagree, from man's perspective we can allow that it is in some sense natural that freethinkers will want their arguments to end up at philosophical naturalism given that philosophical naturalism is the primary prejudice of science. However, even if it natural for science to end up on philosophical naturalism there is still the issue of truth to be considered, because arguments are supposed to be different form assertions.

There never has been a philosopher who did not rely on and learn for experience but we can be sure that philosophers would be concerned about faulty arguments that justified induction and faulty arguments that invalidated induction. In part II of section IV, of his *An Enquiry Concerning Human Understanding* Hume explains that he is arguing independently of his 'faith' in experience: "My practice, you say, refutes my doubts. But you mistake the purport of my question. As an agent, I am quite satisfied in the point; but as a philosopher, who has some share of curiosity I will not say scepticism, I want to learn the foundation of this inference." Regarding the topic of creationism, science cannot argue like Hume because science is convinced that all true arguments must show that creationism is not science.

Although scientific prejudices provide real and ordinary credibility, it is a fact that philosophical naturalism will allow scientists to make the judgement that 'their is no real God who can do things' without looking at the evidence. And, regardless of how good and truthful the theory of evolution is, it is a fact that creationists can legitimately and illegitimately use the primary prejudice of science to dig their heels in against the theory of evolution. As discussed above, science does in fact argue backwards from the truth of the theory of evolution. That there is scope for creationists to legitimately accuse science of thinking backwards from philosophical naturalism may also help to explain why some spokesmen for evolution prefer to argue their confidence with words-and-definitions assertions like 'creationism is not science'.

When people say that 'creationism in not science' they are perhaps thinking that creationism has no credibility just as the fairies in the garden theory has no credibility because we all have credibility at the back of our minds regardless of the words-and-definitions justifications. Philosophical naturalism provides the credibility. Philosophical naturalism is Hume's 'hidden power' that makes sense of the words-and-definitions. We could have the majority of scientists believing in creationism and still conclude that creationism is not testable therefore it is not science. There may be lots of reason why we can say that creationism is not science but we are mostly interested in the reasons that say that creationism has no credibility and therefore it cannot be taught in the science classroom.

We can allow that word categorisations provide a little credibility. If every creationists was a Biblical creationists that might mean that spokesman for evolution could justify a refusal to consider a creationist argument, say, because he is 90% confident that he will be a wasting his time. However if the Bible taught the theory of evolution then these same spokesmen for evolution would not be saying that evolution is not science because it comes from the fixed doctrines of the Bible.

Who hasn't been guilty of using derogatory arguments and words and definitions arguments? Who hasn't been guilty of being deluded by explanatory arguments and prejudices? Science is made up of lots of individuals and therefore there will be lots of mistakes. However I am complaining about the fact that there is no such thing as a scientific voice that is willing to treat errors that end up at the right place as if they were errors that ended up at the wrong place.

What this means is that my real complaint is about the scientific culture not individuals that make mistakes. If a creationists suggests that science is biased against creationism, then what does it matter, because being biased against creationism is like being biased against fairies in the garden.

I am not so much thinking that 150 years have gone by and science has not addressed the natural bias of science, as 150 years have gone by and science has no understanding of the natural bias against creationism. More than 150 years after Darwin published his *Origin of Species* and science still understands itself to be a collection of scientific thinkers, critical thinkers, freethinkers and enlightened thinkers. More than 150 years have gone by and science does not even understand that science has an objective when the topic is evolution? Science has a scientific understanding of creationism and a scientific understanding of truth.

Chapter 5 COMPETING SOLUTIONS

Creationists claim that we can all see creation around us. Because we live in the age of machines, we can call the argument for a creator the design argument or the machine argument.

The theory of evolution claims to offer a viable natural explanation for creation. However, even before the God-is-dead enlightenment freethinkers understood that a natural explanation was the right place for arguments about creation to end up.

David Hume rejected the design argument before Darwin proposed a natural explanation. On p 33 - 34 of *Philosophy of Biology*, Elliot Sober explains why Hume's argument fails. According to Sober, Hume essentially interpreted that the design argument relied on the idea that we are similar to machines (say a watch) and concluded that we aren't all that similar to machines, so the design argument (which Hume interpreted as an inductive argument) was not a very strong argument. Hume also concluded that if we believe that we are constructed by an intelligent designer, then we would have to base this conclusion on experience we have with intelligent designers of other worlds (which is more inductive reasoning). Sober explains that we do not need to use inductive reasoning because we can reason from the best explanation.

Hume may have thought that induction was the only source of empirical knowledge, but nevertheless, it should have been obvious to Hume that 'creation does suggest a creator' regardless of the fact that we are not very similar to watches and regardless of the fact that we have not seen supernatural creators at work and regardless of the fact that he could not find a fault in his argument.

Regarding the problem of induction, Hume managed to argue independently of what he 'knew' to be the truth. However regarding the problem of creation, Hume allowed the fact that he saw something very unacceptable in God-solutions to determine the right place for his argument to end up. Instead of Hume understanding that philosophical naturalism is what makes God-solutions unacceptable to freethinkers, Hume seems convinced that there must be some other explanation.

Sometime after Hume argued against the design argument, William Paley (1743-1805) developed the design argument in his *Natural Theology.* The following citation from Paley's *Natural Theology* has been taken from the Wikipedia article on the watchmaker analogy:

> In crossing a heath, suppose I pitched my foot against a stone, and were asked how the stone came to be there; I might possibly answer, that, for anything I knew to the contrary, it had lain there forever: nor would it perhaps be very easy to show the absurdity of this answer. But suppose I had found a watch upon the ground, and it should be inquired how the watch happened to be in that place; I should hardly think of the answer I had before given, that for anything I knew, the watch might have always been there... There must have existed, at some time, and at some place or other, an artificer or artificers, who formed [the watch] for the purpose which we find it actually to answer; who comprehended its construction, and designed its use... Every indication of contrivance, every manifestation of design, which existed in the watch, exists in the works of nature; with the difference, on the side of nature, of being greater or more, and that in a degree which exceeds all computation.

According the same Wikipedia article "the watch analogy (step 1) does not function as a premise to an argument — rather it functions as a rhetorical device and a preamble. Its purpose is to establish the plausibility of the general premise: *you can tell, simply by looking at something, whether or not it was the product of intelligent design.* In most formulations of the argument, the characteristic that indicates intelligent design is left implicit. In some formulations, the characteristic is orderliness or complexity (which is a form of order). In other cases it is *clearly being designed for a purpose,* where *clearly* is usually left undefined."

We have no choice but to agree that Paley's argument is a rock solid argument for mechanical machines. If the author of this Wikipedia article found a machine in some old rocks, we can be sure that he would believe that the machine was created by a creator, regardless that he knew with extreme scientific confidence that there were no creators around to create.

Spokesmen for evolution would like to remind readers that, as it turns out, Paley's argument is not a good argument for biochemical machines because biochemical machines can reproduce. Whether or not it is reasonable for modern science to suppose that the 'ability of biochemical machines to reproduce solves the machine argument' is something that must be discussed, but whatever the case we should be able to agree that there is nothing rhetorical in Paley's argument. This is to say that even given the first single-celled biochemical machine that could reproduce, then except for the prejudice of philosophical naturalism, the assumption that 'there could be one or two obstacles in getting from such a machine to a walking, talking, self-maintaining multi-cellular biochemical machine without the help of an intelligent designer' is a good assumption. And even if we suppose that evolution can be solved by the 'ability of biochemical machines to reproduce' we still need to get from nothing to a biochemical machine that cannot reproduce and then to one that can reproduce. And again except for the prejudice of philosophical naturalism, we should assume that could be a problem.

The prejudice of philosophical naturalism is similar to the prejudice of repeatability. Some people may insist that it was reasonable for Hume to be very confident that philosophical naturalism held the answer. However, if we are simply going to use philosophical naturalism to decide where we want our arguments to end up then why bother arguing?

Hume was ahead of his time because he believed in some kind of natural solution before Darwin paved the way. However, if we brought Hume into the enlightened age, then my guess is that he would understand the need for an explanation because the right place can change with the circumstances. Proverbs 20:14 says: " "It's no good, it's no good!" says the buyer; then off he goes and boasts about his purchase". I like to call this kind of truth 'business truth'. Business truth is about arguments that end up at the right place. This proverb has the right place changing with the circumstances.

In the modern world the right place has been nailed down by science. Even the fact that Sober is happy to criticise Hume's naturalism is no exception to the understanding that freethinkers want their arguments to end up in the right place, because Hume's argument side-steps the need for an explanation, whereas the modern theory of evolution recognises the need for an explanation.

Like Paley, the Bible also has the understanding that 'the idea that there is a real creator' is the manifest truth of creation that we all see around us. However, in the interest of being able to get a discussion going, it is better to say that a creator is a contender for the best explanation. A best-explanation-theory is established to the degree that the alternative explanations are either eliminated or are judged to be improbable.

Science and religion overlap on the problem of creation. The success of the evolutionary world view requires the failure of the Biblical world view that 'there is a real intervening God' and the success of the Biblical world view requires the failure of the scientific world view that 'naturalism is all there is'. Intelligent design and the theory of evolution are competing best explanation theories.

Sober's 'Gremlins in the attic theory' discussed in the introduction to the previous chapter tells us that best explanations are about prior probabilities. Paley's machine argument assures us that best explanation thinking can be extremely reliable if used properly. Best explanation thinking tells us that we can be certain that the fossils we find in old rocks are the remnants of living creatures rather than the remnants of some natural process. Natural processes can create crystal structures, but we can be certain that the fossils are the remnants of living creatures, because best explanation thinking can be extremely reliable thinking.

In theory, we are interested in the 'prior' probability that there is a God — 'prior' being before we notice that we need a creator to explain creation. Alternatively, we could say that 'prior probability' means 'the probability that there is a God assuming that we do not need a creator to explain creation'.

If we found a watch in some old rocks, then in theory, we should be interested in the 'prior' probability that there were intelligent designers around to make Paley's watch, except that we know there are no alternative explanations. This should mean that providing we restrict our argument to mechanical machines that cannot reproduce, we all understand that there is no such thing a scientific certainty that is more reliable than Paley's argument.

The popular 'evolution is not anti-religion claim' suggests that there are lots of people who agree that the idea that 'there is a god' has a significant prior probability independent of the creation factor. Although the creation factor is not the only religion factor, there is the question of whether these other religion factors are truly independent of the creation factor. We can sidestep this complexity if we can eliminate a natural explanation.

Darwin was once persuaded by Paley's design argument because the argument is simple and obvious. Darwin eventually rejected intelligent design and argued that there is a natural explanation for the purpose that we all see around us. According to Darwin's thinking the fact that biochemical machines we see around us can reproduce allows for a natural solution to the problem of creation of biochemical complexity.

5.1 Intelligent Design

Life has order and therefore it requires an explanation. Here the word 'order' is similar to 'complexity' but in other circumstances, it is more like 'simplicity'. If we are interested in the specificity of arrangement, then chaos is complex. We could say that a lump of dirt is complex providing we have an interest in the precise description of the lump of dirt. Normally we would not be interested in such specificity. Normally one lump of dirt is like any other lump of dirt.

We say that the surface of the water is flat where 'flat' describes the order in the surface of the water. Compared to the surface of flat water, we might feel that the snow flake pattern is complex, however, because it makes sense to define simplicity as an arrangement that predicts itself, then all patterns are really examples of simplicity. A small part of the snow flake pattern predicts the rest of the snow flake pattern. In other words we can say that we have order because we have sufficient bits of raw information to discover the order and confirm the order and then make predictions. For this to be possible, we must be able to discover and compact the description of the shape of the snowflake into a simple rule.

Although the snowflake is an example of simplicity, we can say that pattern of the snowflake is more complex than the pattern of the surface of still water because we need to know more about the snow flake than we do about the water before we know everything.

Simplicity can also be a matter of perspective. Whether or not something is determined to be 'simple' depends on whether or not we have discovered the rule (or seen the pattern) that reduces the amount of information needed to describe the information. Before we have a rule for compacting the information, we could say that we have 'raw information' and if the information can be compacted, we can say that it is 'reducible information' and after we have a way of compacting the information, we could say that we have 'compact information'.

Having discovered and confirmed empirical order, if the amount of order is sufficient we say that the order needs an explanation.

When explaining order, we can allow empirical solutions to the problem of creation of order, if the order is associated with empirical properties of the substances or processes. We can allow that substances organise themselves providing the order is associated with the properties of the substances. The simplicity of the surface of still water and of a crystals for example can be explained in terms of empirical properties.

According to a Wikipedia article titled *Teleological Argument*, Aristotle argued that 'knowledge of what use an animal makes of a feature is crucial to understanding it (for example, that birds use wings for flight)'. This argument suggests that instead of using the design order to predict the performance specification we can observe what the machine does to determine the performance specification and then use the performance specification to predict the design order. In other words, instead of saying that the order of life can be compacted into a simple pattern, we can say that the order of life can be compacted into a simple performance specification.

A statement of the purpose of the machine is a way of compacting the design order because the purpose of the machine is really a performance specification for the machine. The essential feature of the design problem is that it requires the creation of complex order that is related to the function or purpose of the machine. Or working backwards we can say that the purpose of the machine predicts the design of the machine.

It makes sense to say that machines can do things and, therefore, there is purpose in the design of machines, except that people sometimes have a problem with the word 'purpose' because purpose implies an agent. Perhaps there is some rhetorical advantage in the word purpose, but nevertheless it seems impossible not to notice the purpose in the design of biochemical machines regardless of whether we believe that the machines were created by a creator. In *Physics* Book II, section 8 Aristotle argued that there was purpose in the design of biochemical machines even though he believed that creation was not the work of a creator:

> A difficulty presents itself: why should not nature work, not for the sake of something, nor because it is better so, but just as the sky rains, not in order to make the corn grow, but of necessity? What is drawn up must cool, and what has been cooled must become water and descend, the result of this being that the corn grows. Similarly if a man's crop is spoiled on the threshing-floor, the rain did not fall for the sake of this — in order that the crop might be spoiled — but that result just followed. Why then should it not be the same with the parts in nature, e.g. that our teeth should come up of necessity — the front teeth sharp, fitted for tearing, the molars broad and useful for grinding down the food — since they did not arise for this end, but it was merely a coincident result; and so with all other parts in which we suppose that there is purpose? Wherever then all the parts came about just what they would have been if they had come be for an end, such things survived, being organized spontaneously in a fitting way; whereas those which grew otherwise

> perished and continue to perish, as Empedocles says his 'man-faced oxprogeny' did.
>
> Such are the arguments (and others of the kind) which may cause difficulty on this point. Yet it is impossible that this should be the true view...
>
> The end and the means towards it may come about by chance. We say, for instance, that a stranger has come by chance, paid the ransom, and gone away, when he does so as if he had come for that purpose, though it was not for that that he came. This is incidental, for chance is an incidental cause, as I remarked before. But when an event takes place always or for the most part, it is not incidental or by chance. In natural products the sequence is invariable, if there is no impediment.
>
> It is absurd to suppose that purpose is not present because we do not observe the agent deliberating.

Aristotle used purpose to eliminate the possibility of Empedocles' chance explaining life. Empedocles lived from about 490 to 430 BC but we can see both philosophical naturalism and Darwin's natural selection algorithm working in Empedocles' solution. Creation suggest an agent, design suggests an agent, and purpose suggests an agent. There may be some rhetorical advantage to the word purpose but we know that purpose is evident in biochemical machines because we can use the purpose of the machine to compact the description of the machine.

Creationists argue that only an intelligent designer can explain the purpose of biochemical machines because the design associated with the purpose is too complex to imagine that there could be a solution without an agent. Both the modern theory of evolution and Empedocles' theory of evolution claim that natural selection can explain the creation of the complex arrangement of parts that relate to the purpose of the machine.

Freethinkers find something very unacceptable in God-solutions because the idea that a supernatural God intervened conflicts with philosophical naturalism. It seems that because philosophical naturalism insists that philosophical naturalism is the truth, freethinkers feel compelled to find different ways of understanding and communicating their sensitivity to God-solutions. Sometimes freethinkers say things like 'creationism is not science'. On page 52 of *The God Delusion* and on page 316 of *The Blind Watchmaker* Dawkins argues that we should reject God-solutions because God-solutions are not closed explanations:

> I am not attacking the particular qualities of Yahweh, or Jesus, or Allah, or any other specific god such as Baal, Zeus or Wotan. Instead I shall define the God Hypothesis more defensibly: *there exists a superhuman, supernatural intelligence who deliberately designed and created the universe and everything in it, including us.* This book will advocate an alternative view: *any creative intelligence, of sufficient complexity to design anything, comes into existence only as the end product of an extended process of gradual evolution.* Creative intelligences, being evolved, necessarily arrive late in the universe, and therefore cannot be responsible for designing it. God, in the sense defined, is a delusion; and, as later chapters will show, a pernicious delusion...
>
> All that we can say about such beliefs [that invoke a creator to explain creation] is, firstly, that they are superfluous and secondly, that they *assume* the existence of the main thing we want to *explain*, namely organised complexity. The one thing that makes evolution such a neat theory is that it explains how organised complexity can arise out of primeval simplicity. If we want to postulate a deity

> capable of engineering all the organized complexity in the world, either instantaneously or by guiding evolution, that deity must already have been vastly complex in the first place. The creationist, whether a naive Bible-thumper or an educated bishop, simply postulates an already existing being of prodigious intelligence and complexity.

And on pages 70 to 71 of *Darwin's Dangerous Idea* Dennett (and Hume) indentifies the same problem with creationist arguments:

> Only a theory with the logical shape of Darwin's could *explain* how designed things came to exist, because any other sort of explanation would be a vicious circle or an infinite regress (Dennett 1975)...
>
> ...however, as Hume had noted: if God created and designed all these wonderful things, who created God? Supergod? And who created Supergod? Superdupergod?

Hume sees the need for a God to create God because he finds the Biblical understanding of a very big God to be unacceptable. Like modern science Hume found the idea of God to be unacceptable but he had no problem with the idea of super-duper naturalism.

There is something to be said for closed explanations because closed explanations are based on the familiar and the prior probability of the familiar is 1. It is better if our explanations remove the surprise and the unknown so that we have a predictable situation (or rather a retrospectively predictable situation). Hume's 'who made God' argument is saying that the God argument raises other issues that in turn need explaining.

Aristotle's theory, that creation always existed, ends in another mystery. And Aristotle's theory that 'purpose is a property of our world' ends in another mystery. Hume, Dawkins, and

Dennett want to stick to the familiar. Intelligent design is an open explanation that leads to other questions whereas the theory of evolution is a closed explanation that does not lead to other questions.

Although Russell sees something unacceptable about God solutions, Russell disagrees with Hume, Dawkins and Dennett regarding the need for closed explanations. In explaining the ancient Greek philosophy of atomism founded by Leucippus and Democritus (~460~370 BC) and adopted by Epicurus (~341~270 BC) and Lucretius (99-55 BC) Russell argues that theories can be open explanations:

> It was common in antiquity to reproach the atomists with attributing everything to chance. They were, on the contrary, strict determinists, who believed that everything happens in accordance with natural laws. Democritus explicitly denied that anything can happen by chance.! Leucippus, though his existence is questioned, is known to have said one thing: 'Naught happens for nothing, but everything from a ground and of necessity.' It is true that he gave no reason why the world should originally have been as it was; this, perhaps, might have been attributed to chance. But when once the world existed, its further development was unalterably fixed by mechanical principles. Aristotle and others reproached him and Democritus for not accounting for the original motion of the atoms, but in this the atomists were more scientific than their critics. Causation must start from something, and wherever it starts no cause can be assigned for the initial datum. The world may be attributed to a Creator, but even then the Creator Himself is unaccounted for. The theory of the atomists, in fact, was more nearly that of modern science than any other theory propounded in antiquity...

> In regard to both questions alike, there is a limitation which is often ignored, both in popular thought and in philosophy. Neither question can be asked intelligibly about reality as a whole (including God), but only about parts of it. As regards the teleological explanation, it usually arrives, before long, at a Creator, or at least an Artificer, whose purposes are realized in the course of nature. But if a man is so obstinately teleological as to continue to ask what purpose is served by the Creator, it becomes obvious that his question is impious. It is, moreover, unmeaning, since, to make it significant, we should have to suppose the Creator created by some super-Creator whose purposes He served. The conception of purpose, therefore, is only applicable within reality, not to reality as a whole.
>
> A not dissimilar argument applies to mechanistic explanations.
>
> One event is caused by another, the other by a third, and so on. But if we ask for a cause of the whole, we are driven again to the Creator, who must Himself be uncaused. All causal explanations, therefore must have an arbitrary beginning. That is why it is no defect in the theory of the atomists to have left the original movements of the atoms unaccounted for.

Hume, Dawkins and Dennett have overstated the need for closed explanations. Newton's science was criticised because Newton did not explain the cause of gravity. Quantum mechanics seems to lead to other questions but no one seems to be bothered about quantum mechanics. Although Hume, Dawkins and Dennett are pushing the argument for closed explanations too far we can agree that closed explanations are superior because they only refer to the familiar.

If we can reduce our understanding to familiar principles, then we will remove the surprise and the unknown. If we have a predictable situation, then the mystery goes away and in that case we may even find the need to remind ourselves that we do not have any real understanding of the situation. Even though we do not have any understanding of force-by-body-contact, most people would be happy to call such explanations closed explanations, because we need body contact forces to explain so many things.

In our modern world we think in terms of philosophical naturalism versus divine intervention. Before the mystery of our natural world was fully unravelled by the scientific revolution it was the case that people (Hobbes at least) thought in terms of 'miracles' and 'familiarity'. Thomas Hobbes (1588-1679) writes in his *LEVIATHAN* (1651, 36 years before Newton's Principia) part 3, chapter 37 *OF MIRACLES AND THEIR USE:*

> To understand therefore what is a miracle, we must first understand what works they are which men wonder at and call admirable. And there be but two things which make men wonder at any event: the one is if it be strange, that is to say, such as the like of it hath never or very rarely been produced; the other is if when it is produced, we cannot imagine it to have been done by natural means, but only by the immediate hand of God. But when we see some possible natural cause of it, how rarely soever the like has been done; or if the like have been often done, how impossible soever it be to imagine a natural means thereof, we no more wonder, nor esteem it for a miracle.
>
> Therefore, if a horse or cow should speak, it were a miracle, because both the thing is strange and the natural cause difficult to imagine; so also were it to see a strange deviation of nature in the production of some new shape of a living creature. But when a man, or other animal,

engenders his like, though we know no more how this is done than the other; yet because it is usual, it is no miracle. In like manner, if a man be metamorphosed into a stone, or into a pillar, it is a miracle, because strange; but if a piece of wood be so changed, because we see it often it is no miracle: and yet we know no more by what operation of God the one is brought to pass than the other.

The first rainbow that was seen in the world was a miracle, because the first, and consequently strange, and served for a sign from God, placed in heaven to assure His people there should be no more a universal destruction of the world by water. But at this day, because they are frequent, they are not miracles, neither to them that know their natural causes, nor to them who know them not. Again, there be many rare works produced by the art of man; yet when we know they are done, because thereby we know also the means how they are done, we count them not for miracles, because not wrought by the immediate hand of God, but by mediation of human industry.

Furthermore, seeing admiration and wonder is consequent to the knowledge and experience where with men are endued, some more, some less, it followeth that the same thing may be a miracle to one, and not to another. And thence it is that ignorant and superstitious men make great wonders of those works which other men, knowing to proceed from nature (which is not the immediate, but the ordinary work of God), admire not at all; as when eclipses of the sun and moon have been taken for supernatural works by the common people, when nevertheless there were others could, from their natural causes, have foretold the very hour they should arrive; or, as when a man, by confederacy and secret intelligence, getting knowledge of the private actions of an ignorant, unwary man, thereby tells him what he has done in former time, it seems to him

> a miraculous thing; but amongst wise and cautelous men, such miracles as those cannot easily be done.

Familiarity teaches us that God does not intervene in science, but it does not teach us that the laws of the universe are natural laws. However I am not suggesting that this is an argument for God.

Freethinkers may claim that things we are *familiar* with are not miracles, if that is how we want to define the word miracle. Scientifically ignorant people who do that, may find miracles everywhere and freethinkers will use such errors to assert that there are no miracles because all supposed miracles are merely things that we do not understand yet.

It would be better to define 'miracle' by our level of astonishment in the absence of familiarity. If we take this approach then we will find that many familiar miracles are much more miraculous than Jesus walking on water. If we take this approach then a tree, a bird, or a cat is such a miracle because familiarity with things like body contact forces is not the same as familiarity with biochemical complexity. This is because Paley's argument works for things like biochemical complexity regardless of how familiar we are with such things.

The freethinkers understanding of familiarity namely that 'miracles are those things which we do understand yet' comes from philosophical naturalism and it clashes with common sense. Common sense says that Hobbes example of 'an animal engendering its like', is more miraculous then Joshua commanding the sun to stop moving and it is more miraculous than Jesus walking on water. How can even a very big God create biochemical machines that can reproduce themselves? In understanding this argument we must remember that the reproduction of biochemical machines must operate within the constraints of the natural laws rather than by supernatural intervention. Except that we know we are here it is difficult to believe that even a very big God could make such things.

Regarding the problem of 'an animal engendering its like' Christopher Hitchens writes on page 64 of *God is not Great*: "Aquinas half believed in astrology, and was convinced that the fully formed nucleus (not that he would have known the word as we do) of a human being was contained inside each individual sperm. One can only mourn over the dismal and stupid lectures on sexual continence that we might have been spared if this nonsense had been exposed earlier than it was." It was during Hobbes' lifetime in 1677, four centuries after Thomas of Aquinas (1225-1274) that the microscope was invented and people claimed that they had seen miniature humans inside the sperm. If we find that hard to understand then that is because we have lost some perspective on the miracle of 'an animal engendering its like'.

One reason why we may have lost some perspective on the miracle of 'an animal engendering its like' is that we no longer wonder about the details because the solution is no longer in the realm of conjecture, just as we do not wonder about how a Jumbo Jet can fly because we know that it can fly. Instead we use a kind of faith to assert that the details are true.

The modern solution requires that after the original ovum has been fertilized by the sperm the combined cell reproduces. As the cell reproduces some cells to turn into nerve cells some to turn into skin, some into bone, arteries, muscles, etc, and such changes happen in perfect order with the skin on the outside and the eyes in the right place, etc. Not that any of us can begin to appreciate the complexity of how an animal engenders it's like but if we consider the details then we may be able to understand why a few scientists not so long ago thought that the sperm cells had little men in them. Even if the theory of evolution is true, we should be able to understand why it makes sense to say that 'Joshua commanding the sun to stop moving does not compare with the miracle of an animal engendering its like'. If we put aside our familiarity and confidence then we should be able to understand that the the modern solution to how an 'animal engenders its like' it is utterly absurd (utterly absurd but true) and we should be able to understand why people thought that the sperm had little men in them. Because we know that we are here some of us believe in a very big God and some of us believe in super-duper naturalism.

Although Hume, Dawkins, and Dennett have overstated the need for closed explanations, we can agree that compared to Darwin's theory, intelligent-design is a very open theory. Artificial selection can direct changes and we know that there must be such a thing as natural selection. Put these two things together (and providing we go along with Darwin's assertion that the variation achievable by artificial selection can go on forever), then Darwin's argument was a predictive argument, because given life and Darwin's assertion, we must get evolution in the direction of natural selection.

The problem with Darwin's argument was that it included the freethinkers understanding of familiarity. Prior to Darwin, the fact that people were familiar with the absurd idea that wheat can grow from a wheat grain, means that people gave credence to the idea that mice can grow from wheat grains. This was not bad science because mice growing from wheat grains is really only a minor addition to a familiar absurdity. And why not maggots from meat, given that we can get fully formed cattle from a single cell. However, frogs from mud was something completely different. Frogs from mud required a large dose of the freethinkers understanding of familiarity, because frogs from mud clashes with common sense. It wasn't until 1859, (the year that Darwin published his Origin of Species) that Louis Pasteur's experiment finally won the argument against spontaneous generation.

The issue here is that familiarity with slow changes teaches freethinkers that slow changes are natural. Darwin's argument, like spontaneous generation, included some natural magic. This was possible because Darwin was familiar with the absurd idea that wheat can grow from a wheat grain.

That Darwin's explanation seemed to only refer to the familiar, was what made Darwin's argument such a strong argument and it is what made Darwin's argument fundamentally a wrong argument. It was wrong on a theoretical basis because without some natural magic it would be impossible to have his sort of evolution unless all the necessary design information was already represented in the genes. Wheat can grow from a wheat grain because the wheat grain has all the necessary wheat-information to tell it how to grow. If we want a fish to evolve to man then the fish has to have all the man-information.

People who believe in the theory of evolution will insist that Darwin was on the right track. However, even if it somehow turned out that spontaneous generation worked then although spontaneous generation happened to be on the right track, we should not say that it was on the right track unless we believe that scientific insanity can be on the right track. Spontaneous generation did not allow for the generation of the design information. Spontaneous generation effectively assumed that there was some law of the universe that designed machinery.

We can say that Newton's law of gravity is an open explanation or we can say that it is an empirical explanation. However, given Newton's laws and some understanding of the structure of the moon, we expect the explanation for why we only see one face of the moon to be a closed explanation. Closed explanations are predictive explanations because a closed explanation is really a way of reordering and refocusing our existing knowledge. In other words, a closed explanation is a logical solution rather than an empirical solution.

We can say that Darwin's explanation was an attempt at a closed explanation or it was an attempt at a logical explanation because that is what we must have for a non-supernatural theory that explains life because we know from Paley's argument that there cannot be an empirical law of the universe that designs machinery. An intelligent designer can explain the creation of design order, but an empirical law (such as Darwin's theory, spontaneous generation and ancient Greek atomism) cannot because we cannot have an empirical law that makes machines, because the order of a machine is associated with the purpose of the machine not the properties of the materials. Intelligent design is an open explanation (or an empirical explanation) whereas Darwin's theory of evolution was an attempt at a logical explanation with the hidden natural magic of spontaneous generation.

If Darwin's solution worked as an explanation, then we could say that 'in any case the God explanation was never a good explanation because it is not a closed explanation'. If naturalism works then it is going to be the best explanation by a very long stretch because we are interested in credibility. If that was the case then we could say that as it turns out a closed explanation explains what seems to be an impossible problem. A closed explanation that works is a credible explanation, whereas we only resort to open explanations when there are no closed explanations.

Modern science has modified Darwin's theory to allow it to generate new design information because Darwin's argument was flawed. Darwin should have been wondering where the design information came from. Instead of saying that Darwin's argument was a flawed argument, we could say that it was on open explanation that looked like a closed explanation. For Darwin's argument to be a closed explanation, it would need to explain how it was that a fish carried all the necessary man-information.

Although Dawkins argues that the real question is the question of explaining organised complexity (that relates to purpose of the machine) the issue is not so much that complexity is improbable but that some kinds of complexity are improbable if naturalism is all there is. Dawkins book for example is organised complexity but it is very easy to explain. When it comes to explaining life we need either a logical explanation or we need an intelligent designer. To insist that the real question is the question of explaining organised complexity is to make naturalism an unbeatable system.

Freethinkers find something unacceptable about God solutions but they are not able to put their finger on exactly what it is so unacceptable. This something is the gravity philosophical naturalism. It is a gravity because it works on our thinking without our knowledge. The gravity of philosophical naturalism has an aversion to the God word. The big bang theory leads to other questions but freethinkers like the big bang theory because the big bang theory avoids the God word.

Darwin's solution was acceptable to science because philosophical naturalism is the primary prejudice of science. In other words Darwin's theory was acceptable to science for the same reason that Russell argues that the ancient Greek theory of atomism and the ancient Greek theory of evolution were more scientific. Alternatively we could say that the concept of familiarity helps freethinkers to believe that miracles are things we do not understand yet. Familiarity with things like wheat growing from a wheat grain helped freethinkers to ignore the natural magic in Darwin's theory of evolution, the magic in spontaneous generation and the magic in ancient Greek science. To grow slowly seems natural because we are familiar with things that grow slowly.

Because we are interested in credibility and the justification for intelligent design is based on elimination of contending theories, it may not matter if the God explanation is a weak explanation or if it invokes the greatest fudge factor of all time or if it cuts the Gordian knot or if it is not a complete explanation. All these things count towards credibility but what we should care about is whether on an overall basis it is reasonable to believe in intelligent design.

An aversion to the God word means that freethinkers are frustrated by the understanding that intelligent design cuts the Gordian knot while being completely insensitive to the fact that if naturalism can design self-replicating, self-maintaining biological machines and a universe to put them in, then science has replaced the idea of an omnipotent God with super-duper naturalism.

Fortunately, we are only familiar with things that grow slowly. This means we should also be able to say that naturalism can't solve the problem of the first life form, except that the freethinking bug forbids saying such things. According to an NCSE article titled Creationism and Pseudomathematics:

> We are well aware of anti-evolutionists' fondness for presenting their audiences with numbers of dizzying magnitude that they use to represent incredibly low probabilities for such events as the chance formation of a protein molecule, the origin of life, and the like. Thus they argue that it is irrational to believe that the event in question could have happened naturally (they mean "by chance") without the aid of intelligent design. In some cases, such as the chance formation of habitable planets, one may avoid a technical discussion of the physical

> processes involved and respond simply by pointing out that the universe is a very big place, containing countless galaxies, stars, and planetary systems, thus providing so much opportunity for the natural occurrence of the event in question that the probability may be quite high that such an event would occur somewhere. Furthermore, if the universe is infinite, providing the event with infinitely many chances to occur, then the occurrence of the event is a virtual certainty. Thus creationist probability arguments can often be undermined by pointing out that any event with a probability greater than 0, no matter how low, will be likely to happen if given enough opportunity, and sure to happen if opportunity is unlimited....
>
> **ALL NONZERO PROBABILITIES ARE POSSIBLE**
>
> The probability of an event is expressed as a real number from 0 to 1; the more probable the event, the higher the number. An event can have only one probability at any time, just as a person at any given time can have only one age. However, anti-evolutionists misconstrue Borel's law of chance to imply the absurdity that low-probability events are assigned 2 different probabilities - their *true* probability and a probability of 0.

The reason for the dizzying magnitude of such large numbers is that to assume a simple bacteria came into existence from nothing is a stretch. Unless we believe in things like spontaneous generation, it seems reasonable to assume that the chances of getting a bacteria from nothing are similar to the chances of a complicated mechanical machine being assembled from nothing. This means that even if science assumes an evolution solution following the creation of the first life form, science already understands that the creation of the first life is comparable to finding a machine in some old rocks. Given that there is life on earth and philosophical naturalism then there must be a chance that the universe is infinite.

An infinite universe would mean that there is another planet where all the people are the same as the people on this earth who are living in houses the same as houses on this earth in countries the same as the countries on earth. When you go out to get the mail from your mail box walking on your hands and singing Waltzing Matilda, then there is someone who looks like you in another universe doing the same thing. In fact there are lots of people doing almost exactly the same thing. This is because when we change from very large number (no matter how large) to infinite then we have changed concepts, because infinite is not a size.

An infinite universe seems like a very big mystery to me. However it is a scientific mystery because like the big bang theory it avoids the God word. If our universe is infinite then presumably evolution to a man in one step would be possible on some other planet that we could never find.

Assuming a finite universe and assuming that the earth had a beginning then the creationists claim 'that it is manifestly impossible to have a natural solution to the problem of the first life form', should be easy to understand. If we found a machine in very old rocks, we would explain the find by postulating a creator despite the fact that this raises other questions and even despite the fact we were scientifically certain that no intelligent life existed on earth prior to us — that is we not only accept open explanations but we accept open explanations that contradict the familiar. If science was confident that the universe was infinite, then perhaps science would be happy to agree with creationist claims about the first life form because in that scenario such claims would not take science to the God place.

5.2 The Evolution Solution

Spokesmen for evolution like to ignore the problem of the first life form and remind readers that biochemical machines are different from manmade machines because they can reproduce.

The fact that biochemical machines can reproduce, means that biochemical machines are different from Paley's man made machines. The fact that biochemical machines can reproduce, means that natural selection can work on biochemical machines.

Given that life can reproduce but watches can't, Darwin argued that natural selection could create. Natural selection has long been demonstrated by the power of artificial selection and the logical idea that some designs are more suited to a particular environment than others.

On page 153 of his *Origin Species* Darwin reasoned that because artificial selection can produce significant changes in a short period of time, that much larger changes could be produced in a much longer period of time:

> Slow though the process of selection may be, if feeble man can do much by his powers of artificial selection, I can see no limit to the amount of change, to the beauty and infinite complexity of the coadaptions between all organic beings, one with another and with their physical conditions of life, which may be effected in the long course of time by nature's power of selection.

From a theoretical perspective (even if we allow that in practice there are no limits on selective breeding), Darwin should have wondered where all the information comes from? It cannot come from natural selection if there are no failures, unless we believe that all possible designs work. If all possible designs work then that means that the information was pre-existing just as the information that allows wheat to grow from a wheat seed is pre-existing.

We know that some of the kind of information that Darwin required was available to natural selection, because we know for sure that artificial selection works. Artificial selection is what allows animal breeders to breed an animal to develop some chosen characteristic.

Although life contains some of the sort of information that Darwin required for his evolution solution, animal breeders of Darwin's time insisted that after a number of generations the available variation for a particular characteristic would run out and progress would be halted. Animal breeders insisted that the experimental evidence was that there were fixed limits on the amount of achievable variation. On pages 71 and 72 of *Origin of Species*, Darwin accepted that it did seem as if there were fixed limits because he argued that "organic beings must be exposed during several generations to the new conditions of life to cause any appreciable amount of variation" and that "the most frequent cause of variability may be attributed to the male and female reproductive elements having been affected prior to the act of conception". However on page 77 Darwin argued that the idea of fixed limits was a mistake:

> But there is not a shadow of evidence in favour of this view: to assert that we could not breed our cart and race-horses, long and short-horned cattle, and poultry of various breeds, and esculent vegetables, for an almost infinite number of generations, would be opposed to all experience.

It may be that Darwin confused general hind-sight changes with changes that were the true prior targets of the breeding programmes? Perhaps he thought that animal breeders were being influenced by creationists? Whatever the reason, Darwin insisted that the idea of fixed limits was opposed to all experience, and that is not true.

Scientific thinking insists that it makes sense that Darwin worked backwards from the assumption that there was a natural solution because that is how we use prejudices in our thinking. What happened to Darwin's scientific prejudice against magic? Familiarity and philosophical naturalism happened. Philosophical naturalism provided the gravity. Familiarity with things growing helped Darwin to ignore the mystery of where all the information comes from and assert that changes could continue forever just as familiarity with things growing helped people to believe in spontaneous generation.

If Darwin's assertion was true, then the argument would be over and in fact we could say that large scale evolution by natural selection would be a predicted outcome even without any evidence of large scale evolution, because we would have a closed solution. And if this was the case, the theory of evolution would not need mutations.

Around the end of the 19th century, science recognised the natural magic in Darwin's solution and added a mutation component (and trial and error) to Darwin's theory. Modern evolutionists agree that, contrary to Darwin's assertion, that the availability of changes from existing DNA must come to a halt. Modern evolutionists added a mutation component to allow evolution to go on forever. Modern science now insists that biological complexity can be generated by the following three link logical algorithm:

- a random change component (mutations)
- a natural selection component
- a copy component

The modern theory includes a mutation component. To say that this is a logical explanation is to say that the modern theory of evolution is also a closed explanation rather than empirical explanation.

In the following discussions, when I use the term 'evolution', I mean 'the three link process or algorithm' as distinct from 'the slow and gradual changes that are achieved by natural means'. The modern theory of evolution says that there is a logical solution to the creation of complexity. Effectively it is said that this three link algorithm can create a special kind of order that relates to the selection criterion. And if the selection criterion relates to the ability to survive and reproduce, we will have the beginnings of a solution to the problem of creation of design order by natural processes. This evolution solution is essentially a trial-and-error process with a natural feedback element. Of course, the solution is expressed empirically and at some level it relies on the empirical properties of materials, but the three link scheme that generates the solution is a logical scheme. Any natural solution must be a logical solution because Paley's argument tells us that there can never be a law of the universe that creates life.

No doubt we must have natural selection, given that there are naturally occurring variations. And we know we have random changes (mutations) and the copy process (reproduction), so we know that all three links exist. Of course, we are not really interested in the destruction of complexity (turning a gene off) to create variation although that must happen. For the purpose of investigating the theory of evolution, we are interested in the creation of new useful information.

There are no logical problems with the mutation-copy-selection process. It seems to be that something similar to the evolution algorithm allows our anti-bodies to evolve to select the shape that represents a particular type of foreign object, and that seems to be an obvious application for the evolution algorithm in biological design.

Just as there is no choice but to allow that the selection-copy algorithm (without the mutation) can provide natural selection and there is no choice but to allow that the mutation-selection-copy algorithm can create simple useful changes by turning off genes. Sickle cell anaemia in Africans is an example of this three link process creating new useful simple design information. Sickle cell anaemia is a genetic disease of the blood that provides Africans with some protection against malaria. Because sickle cell anaemia also kills people, most people would agree that this variation was not created by God and therefore it was created by evolution. This means that for the purpose of investigating the theory of evolution, instead of saying that we are interested in the creation of new useful information, we should say that we are interested in new useful information that is complex.

Darwin's solution allowed for the selection of complex designs but it did not allow for the creation of complex designs, and that is essentially why modern science has abandoned Darwinism and moved on to the modern theory of evolution. However, some spokesmen for evolution have not moved on with modern science. Some spokesmen for evolution still use the success of natural selection (the easy bit) to demonstrate the strength of evolution. For example Dawkins argues, from p 40 of *The Blind Watchmaker*:

> Eyes don't fossilize, so we don't know how long our type of eye took to evolve its present complexity and perfection from nothing, but the time available is several hundred million years. Think, by way of comparison,

> of the change that man has wrought in a much shorter time by genetic selection of dogs. In a few hundreds, or at most thousands, of years we have gone from wolf to Pekinese, Bulldog, Chihuahua and Saint Bernard. Ah, but they are still *dogs* aren't they? They haven't turned into a different '*kind*' of animal? Yes if it comforts you to play with words like that, you can call them dogs. But just think of the time involved. Let's represent the total time it took to evolve all these breeds of dog from a wolf, by one ordinary walking pace. Then on the same scale, how far would you have to walk, in order to get back to Lucy and her kind, the earliest human fossils that unequivocally walked upright? The answer is about 2 miles.

If we think that these changes could have happened in a few hundred years, then except for a slight complication, we must be talking about recombination and not creative mutational evolution. Recombination may be expressed the same way as mutation — it may be expressed as changes in form, but of course creationist will assume that none of the changes in the dog generated by selective breeding are a result of new information but rather that they are the result of pre-existing design information. That is creationists would assume that natural selection is doing exactly the same thing as artificial selection. Contrary to what Darwin argued, successful artificial selection demonstrates natural selection, it does not demonstrate evolution.

Given the few hundred year assumption, even Dawkins must be assuming that the newly expressed information was pre-existing, although he would believe that the information was put there by evolution sometime in the past. However, to say that the information was pre-existing (regardless of how the information originally got there) and that it was expressed by natural selection is not an argument for evolution or intelligent design otherwise artificial selection of a new form would be an argument for evolution.

Natural selection is an unbreakable link in the chain of ideas that form the theory of evolution. If we believe in artificial selection then we should believe in natural selection.

New design information is design information created by mutation, whereas recombination works by selecting an existing alternative for a particular gene. Recombination is possible because living things typically carry alternatives (alleles) to each gene. This recombination is what gives power to selective breeding and therefore, in some sense, it can be said to be the source of evolution. But if we want to discuss the viability of evolution then recombination is not evolution because in such a context we are only interested in the creation of new complexity. Creationists could assert that the alternative genes (alleles) were put there by God and that is why artificial selection works. Recombination is a part of evolution. Lots of things are a part of evolution, but for purposes of discussing the viability of evolution as the solution to the design problem we need to focus in the creation of new complex information.

Creationists also confuse recombination and mutation when they argue against natural selection. Creationists have objected to the correlation of the change in colour of the famous *peppered moth* with the effects of industrial pollution on the trees. Even if industrial pollution was not the driving factor, creationist should agree that these sorts of things do occur. Natural selection is just a natural form of artificial selection and therefore, natural selection must occur to the extent that certain natural forms can have an advantage against alternative natural forms in natural environments. By arguing against specific examples of natural selection without suitable clarification of the issues, creationists risk an endorsement of the Darwin's natural magic — that variation by natural selection gives us evolution. Variation by natural selection must happen if we believe that there are different forms that can reproduce and pass on their characteristics and that it is possible for a particular form to be more suited to a particular environment. However, to believe that recombination can go on forever is to believe in spontaneous generation. It is to believe that any amount of changes providing they are slow are not miraculous because we are familiar with the fact that wheat can grow from a wheat grain.

Recombination is the expression of existing design information and in this sense it is similar to the growth of a tree from a seed. Recombination is not recombination of bits and pieces of genes but it is recombination of a gene segment between nodes. The difference being that recombination generally does not result in a failed design because the new segment being introduced is already ordered unlike a segment of random DNA made in a laboratory.

If we are arguing against intelligent design, then we should consider that we have alternatives because God did not want everyone to look identical because God wanted blonde and brunette, short and tall, male and female. But the fact that we have such variation and that we can select to emphasise a particular trait does not give any indication of the truth or otherwise of evolution. Of course, Dawkins could assert that none of the changes were put there by God, but in that case, we might as well follow Darwin's logic and deduce that because we know that animal breeders can create progressive changes, we can assume that changes can go on forever.

The creative mutation process is the process that introduces the new information into the pre-existing stockpile of complex genetic information. This is the only bit in question because it is the bit that seems to require an intelligent designer. The number of failures (detrimental or catastrophic mutations) should be an indication of how much real evolution is taking place. Assuming we are adding to a complex design, then for every bit of new information being put there by evolution (as distinct to new information being expressed by natural selection), we should expect to see lots of failures. This means that either Dawkins is assuming that the newly expressed information was pre-existing, or he is assuming that evolution is like the growth of wheat from a wheat grain.

A new allele might be a gene that no longer has any function because of a mutation. In the case of Dawkins' dogs for example, creationists can allow that a simple mutation that might turn off one of the genes for dog hairs with the result that the dog has less hair or shorter hair. A hairy dog might produce a short haired dog even if God did not design short haired dogs and if a short haired dog can be achieved with simple changes then such changes must be possible. From a creationist's perspective, acceptable causes for the variation of the dogs under artificial or natural selection are:

1. Recombination of the original working genes.
2. Recombination of the original working genes + recombination of pre-existing simple-mutations.
3. Recombination of the original working genes + recombination of pre-existing simple-mutations + new simple-mutations.

Creationists can only object to the creation of design complexity by a trial and error algorithm. Creationist can go along with the other bits such as natural selection, and even simple-mutations that might produce say a bald or short haired dog. As with man made machines we can achieve architectural changes to the shape, colour, finish, etc, quite easily but we cannot easily add to the function of the machine.

We cannot easily add to the function of a machine, but we can easily reduce the function of a machine. If we have a radio-transmitter, then we might be able to convert it to either a radio or a transmitter by poking some wire cutters into it and snipping either the speaker wires or the microphone wires. Even the genes for creation of dog hairs might represent considerable complexity, and therefore, it might be easy for a single mutation to destroy this complexity and turn the dog hairs off.

However, this does not mean that we need even new simple-mutations to explain the variation produced in Dawkins' dogs. Although creationists must accept that mutations can produce some sort of variation, we need evidence to suggest that the variation in Dawkins' dogs expressed by natural selection was generated by simple-mutations. The changes could have been achieved by recombination of the original alleles only, which is to say either causes 1 or 2.

How far have these dogs moved? With respect to creative evolution, there is no evidence to suggest that they have moved at all and for all we know, we are comparing a distance of zero with two miles because Dawkins' example of evolving dogs is really an example of 19th century Darwinism. Unless these changes occurred in combination of numerous failures, then Dawkins should be convinced that evolution of design complexity has not happened.

Even if a dog does happen to turn into a elephant without lots and lots of unsuccessful dog-elephants, then such and example would not be evidence of evolution but it would merely be an example of natural selection. Evolutionists who use natural selection to argue evolution and creationists who argue against natural selection are both a part of this spontaneous generation problem. The issue is that we are too familiar with the absurd idea that wheat can grow from wheat grains.

Natural selection like artificial selection, must happen because genes are represented by more than one allele.

One complication to the understanding that recombination is nothing like a mutation is that we do occasionally get incompatible recombinations such as Rh incompatibility in mother and child. Another complication to the overall argument is that creationists have no choice but to allow for some mutational changes, because mutations do occur and apparently there are lots of them. SNPs (single nucleotide polymorphisms/variations from the normal) are the most common form of variation in the human genome. According to biblical creationism, one man and one woman could at most only carry 4 alleles for any gene. This means that existing designs carrying more than 4 alleles that were not put there by God and if they were not put there by God, then they were put there by mutation.

Although creationists have no choice but to accept that errors do happen and that such errors must result in some differences, if we are saying that lots of mutations can happen without destruction of complexity, then the extreme complexity must be represented by genes that are more stable than those genes given to frequent mutations (SNPs). The idea here is that if we are adding errors to the design of an extremely complex machine, then we might be able to allow such errors, providing that they are restricted to non-essential information.

Regarding the viability of evolution we need to focus on the difficult bits. Fundamentally intelligent design is reasonable theory because a natural solution to the problem of creation of extreme complexity is a big ask. Conversely, if we are not evolving extreme complexity then simple-mutation evolution can happen. In other words we need the biggest all time fudge factor, namely God, to solve those things that seem to require the biggest all time fudge factor to solve. We may not need God to create mountain ranges but we may need God to create self-maintaining, self-reproducing biochemical machines.

The difficult bit of evolution is the creation of new complex information. The difficult bit of evolution is the bit that seems like it requires a creator. From an evolutionary perspective, the difficult bit of evolution should require lots of failures expressed in terms of unfit progeny because if we are changing something that is complex with random changes, then we should expect to get lots of failures. The question then is how many failures we should expect when we are modifying biochemical machines? This should depend on the complexity of biochemical machine. In order to consider the problem of complexity we first need to consider the meaning of 'complexity' in the context of the design problem.

The design problem is a performance problem. It should be the case that even if we know nothing about how a machine such as how a computer performs a complex task, we all understand that it doesn't matter how many times we randomly join together a bunch of electrical components we are never going to get a computer. We know that the design of a computer is going to require an extremely specific solution. This seems to be what we mean by a complex design.

The haemoglobin molecule is a sphere and a sphere is an improbable design, therefore, we might say that it has some complexity. Suppose we used the chance of generating the haemoglobin molecule by accident to represent the complexity. There are twenty different amino acids to choose from and the beta haemoglobin molecule is a specific sequence of 146 amino acids. The ways of arranging these 146 amino acids is 20^{146} which gives us a probability of approximately equal to $1/10^{190}$ or we might say a complexity of 10^{190}.

Although the haemoglobin number gives us some idea of what could be called the 'prescriptive complexity' associated with the design of the molecule from the point of view of a prescriptive specification, it does not give us a true indication of its 'design complexity', because the design problem is fundamentally a performance problem. Fundamentally the purpose of a machine is to perform a task. And that means we need to take into account that we may be able to use a distorted tetrahedron, octahedron, or cube to represent the first working sphere.

As an estimate of the performance complexity of a machine, we could use the ratio of the number of viable solutions to non-viable solutions given a stockpile of components. The reciprocal of this number then gives us the chances of hitting on a workable performance solution (rather than a particular solution) by accident. This ratio ignores the possibility that there may be more than one simplest solution and it ignores the fact that the stockpile of components would generally include other components that are not required but it allows a first guess assessment. This kind of complexity very easily results in large numbers. This kind of complexity seems to be what we mean by complexity but it is not evolution compatible complexity because we cannot get to this kind of complexity one step at a time.

We can also define complexity in terms of what Professor Michael J. Behe calls irreducibly complexity. A design that is irreducibly complex is a design that needs all its working parts to function. This means that after we discard all the redundant components and strip the design down we could determine the complexity in terms of the number, specificity, and interconnection of the components.

Professor Michael J. Behe has argued in his book *Darwin's Black Box — The Biochemical Challenge to Evolution* that biochemical designs are 'irreducible complexity' and therefore, they must be the creation of an intelligent designer. Behe's argument effectively says that machines consist of a type of order that cannot be constructed by a step-by-step evolutionary process. Behe says that machines often require that we have components A, B, C,... to be present at one time in order for the machine to function and it simply is not possible for an accumulation algorithm to create such order. Something that is irreducibly complex cannot evolve because all the components are required to be operating for the machine to work. On p 22 and 25 of *Science and Creationism*, the National Academy of Sciences comment on Behe's concept of 'irreducible complexity':

> Modern-day intelligent design proponents argue that molecular structures such as DNA, or molecular processes such as the many steps that blood goes through when it clots, are so irreducibly complex that they can function only if all the components are operative at once. Thus, proponents of intelligent design say that these structures and processes could not have evolved in the stepwise mode characteristic of natural selection.
>
> However, structures and processes that are claimed to be "irreducibly" complex typically are not on closer inspection. For example, it is incorrect to assume that a complex structure of biochemical process can function only if all its components are present and functioning as we see them today. Complex biochemical systems can be built up from simpler systems through natural selection. Thus, the "history" of a protein can be traced through simpler organisms. Jawless fish have a simpler hemoglobin than do jawed fish, which in turn have a simpler heamoglobin than mammals.
>
> The evolution of complex molecular systems can occur in several ways. Natural selection can bring together parts of a system for one function at one time and

> then, at a later time, recombine those parts with other systems of components to produce a system that has a different function. Genes can be duplicated, altered, and then amplified through natural selection. The complex biochemical cascade resulting in blood clotting has been explained in this fashion...
>
> Creationism, intelligent design, and other claims of supernatural intervention in the origin of life or of a species are not science because they are not testable by the methods of science.

The National Academy of Sciences are saying that evolution can build complex biochemical systems and then in the next paragraph they are saying that there are ways around the problem of creation of the occasional complex systems like the blood clotting mechanism. This is to say that biological design does not have sufficient 'irreducible complexity' to challenge the evolution solution, and that "structures and processes that are claimed to be "irreducibly" complex typically are not on closer inspection." The fact that the blood clotting mechanism needs an explanation means that the blood clotting mechanism seems to contain irreducible complexity, but apparently insufficient to challenge a natural solution.

The National Academy of Sciences must argue, as they have, because there is no other way of arguing against irreducible complexity because irreducible complexity is not evolution compatible complexity. Saying that 'irreducibly complex systems cannot evolve' is like saying 'systems that cannot evolve, cannot evolve'. One might wonder what all the concern is over.

Two things create the concern. Behe showed that something as simple as a mouse trap is irreducibly complex and cannot evolve. And Behe explains how at the microscopic level, things happen just like they do in our much larger world, rather than by natural magic. The Behe perspective is for example, if a cell wanted to send a molecule from one part of the cell to another part of the cell, the dispatch mechanism might need to put some sort of address on the molecule, arrange for some sort of transport vehicle that can read the address and provide some way to ensure the various gates open, load, and unload the molecule, etc. Before that happens, we would need to make the correct molecule. Behe shows us by example that it is in our thinking to imagine simple solutions if we cannot see the details. He tells us that once upon a time, we imagined things could somehow happen without this sort of mechanical complexity and that once upon a time we imagined that maggots could just 'spontaneously generate' from rotten food just as trees can grow from seeds. According to Behe, the microscope has pressed back these ideas and we now know that maggots are not amorphous lumps of jelly with magical properties. We now know that each cell of the maggot is an extremely complicated machine.

Behe is saying that irreducible complexity is similar to ordinary complexity and that it is different to evolution compatible complexity. What do we all know for certain? What can we all agree on? We all know that Behe's type of complexity must exist and it can produce designs and it can be made by an intelligent designer. And we all know that evolutionary changes can happen after we have a working design because we can change things and the machine will continue to work. However, what we do not know is whether evolution compatible complexity can make complex machines. Scientific thinking assumes that it can because philosophical naturalism assumes that evolution happened. In other words, everyone believes that there is a type of complexity that cannot evolve, but evolutionist assert that evolution-compatible complexity is the sort of complexity we need to explain biochemical machines.

We also know that evolution cannot create the simplest possible solution if the simplest possible solution is complex. If the simplest possible solution to a particular performance problem is not complex then the particular problem is not a good example of complexity.

There may be many features that can be modified by a feedback algorithm. There may be many redundant features in a clock, a radio, or a car that can be eliminated and the design will still work. This is because, regardless of how difficult the performance specification is, we can always add components without causing the design to fail. An irreducibly complex design is just something where we need components A, B, C and D to be all present at one time in order for the design to work, but that does not prevent us from adding a non-essential component. This means that we can modify the design one step at a time, providing we have a working solution. If we are changing a design one small step at a time, then we are enhancing the design. One step at a time means enhancement not design. It does not make sense to say that we can design something one step at a time. However these sorts of arguments do not work to undermine the truth of evolution, because evolutionists have a way of understanding how evolution works.

On pages 47 to 49 of *The Blind Watchmaker*, Richard Dawkins illustrates the nature of evolution compatible complexity in his comparison of the probability of cumulative selection of small random variations versus the probability of a single step random variation. In his example, he considers the time taken for a monkey to type the target phrase: "METHINKS IT IS LIKE A WEASEL" versus the time for a computer to generate the target phrase:

> In order to simulate cumulative selection the computer begins by choosing a random sequence of 28 letters and: It now "breeds from" this random phrase. It duplicates it

> repeatedly, but with a certain chance of random error — "mutation" — in the copying. The computer examines the mutant nonsense phrases, the "progeny" of the original phrase, and chooses the one which, however slightly, most resembles the target phrase. The chosen phrase becomes the next generation phrase which in turn is subject to duplication with some random variation, etc, etc, until the target phrase is generated. Computers are a bit faster at this kind of thing than monkeys, but the difference really isn't significant. What matters is the difference between the time taken by cumulative selection, and the time which the same computer, working flat out at the same rate, would take to reach the target phrase if it were forced to use the other procedure of single step selection: about a million million million million million years. This is more than a million million million times as long as the universe has so far existed. Actually it would be fairer just to say that, in comparison with the time it would take either a monkey or a randomly programmed computer to type our target phrase, the total age of the universe so far is a negligibly small quantity, so small as to be well within the margin of error for this sort of back-of-the-envelope calculation. Whereas the time taken for a computer working randomly but with the constraint of cumulative selection to perform the same task is of the same order as humans ordinarily can understand, between 11 seconds and the time it takes to have lunch. There is a big difference, then between cumulative selection (in which each improvement, however slight, is used as a basis for future building), and the single-step selection (in which each new "try" is a fresh one).

Here, Dawkins has a understanding of biological complexity that is analogous to smooth hills. On page 221 of *Darwin's Black Box*, Behe pointed out that this cumulative selection relies on there being a succession of intermediate states that are all viable regardless of which path is taken and therefore, Dawkins' analogy is inappropriate. In the case of the computer generation of a target phrase the computer program guarantees that each state that is most similar to the target state will be selected and guarantees that all selections lie on a path to the target sentence. On p38 of *Philosophy of Biology*, Elliot Sober says that while Dawkins' process is not a perfect analogy in every respect, it does serve to illustrate an important feature of how evolution proceeds. Although Dawkins' analogy is not perfect (because every state cannot be viable) this is only an inaccuracy rather than a fatal flaw in the models reflection of the evolutionary process. This is because, although the evolutionary process does not require that all states are viable states, it does require a succession of viable states. In this respect, Behe and Sober are not saying different things and the problem is not so much with Dawkins' analogy as it is with the evolution solution itself. What Dawkins' analogy does, is remind us that evolution requires one slight improvement being built upon a previous slight improvement, which is enhancement not design. Obviously it is enhancement because if we are adding one step at a time then we must be adding one step at a time to a working solution. On page 147 of *The God Delusion*, Dawkins argues:

> What is it that makes natural selection succeed as a solution to the problem of improbability, where chance and design both fail at the starting gate? The answer is that natural selection is a cumulative process, which breaks the problem of improbability up into small pieces. Each of the small pieces is slightly improbable, but not prohibitively so. When large numbers of these slightly improbable events are stacked up in series, the end product of the accumulation is very very improbable indeed, improbable enough to be far beyond the reach of chance. It is these end products that form the subjects

> of the creationist's wearisomely recycled argument. The creationist completely misses the point, because he (women should for once not mind being excluded by the pronoun) insists on treating the genesis of statistical improbability as a single, one-off event. He doesn't understand the power of *accumulation*.

Philosophical naturalism insists that biological complexity must be like that because there is life on earth. Evolutionists have in mind a special kind of complexity that is compatible with a logical trial and error solution. People who confuse natural selection for evolution have this understanding of evolution. People who think the fact that 'there are genetic differences and that genetic inheritance occurs and that animals that cannot compete will die' is an argument for evolution are thinking in terms of evolution compatible complexity.

Evolution compatible complexity has an understanding of biological complexity that is like a smooth hill in an otherwise flat landscape. The creative ability of evolution is said to be in the selection because selection determines the direction. And the power of evolution is said to be in the accumulation process — evolution can climb large mountains because any number of steps can be accumulated. This argument works in our imagination because we can imagine an amoeba changing into a different kind of amoeba and so on all the way up to man. But of course when we think like that we have in mind enhancement because we already have a working solution.

Evolutionists have in mind a kind of complexity that can be broken up into small pieces. Evolutionists want evolution to be able to open a sophisticated combination lock one number at a time because evolutionists have in mind enhancement rather than design.

Because the peak of the evolutionary hill represents only a small portion of the total area of the landscape, we might say the peak is in some sense highly improbable. However, the peak is certain give a climbing algorithm and the correct starting position. Evolution is a process that climbs to the top of these design peaks, like a mathematical algorithm finds the peak on a curve which is nothing like design. Each evolutionary step up the hill is a step taken by a progeny relative to the position of the parent. If it makes sense to call what is being created complexity (because the peaks are highly specific positions), then we should call it evolution-compatible-complexity because it is different to ordinary complexity. And if we want to call that design rather than enhancement, we need to call it evolution-compatible-design because it is not what we would normally have in mind for design.

Enhancement is about modifying a machine or an aspect of a machine that works. If we are modifying something that works, then we already have something that works. Ordinary complexity is about the number of non-viable ways of rearranging the simplest possible solution, or it is about the ratio of the number of viable solutions to the number of non-viable solutions. When we have in mind evolution that is like climbing a smooth hill (when every state is a viable state as is the case for Dawkins' computer generated sentence), then the ordinary complexity will be 0. Adding a few pot holes to the enhancement hill will reduce the probability of success and the reciprocal of probability numbers give a measure of ordinary complexity. This seems to satisfy our imagination, providing we do not ask ourselves whether the sorts of probability numbers we have in mind reflect the complexity of biochemical machines. On pages 81 to 82 Christopher Hitchens writes:

Now we are being told that astonishing features, such as the human eye, cannot be the result of, so to speak, "blind" chance. As it happens, the "design" faction have chosen an example that could not be bettered. We now know a great deal about the eye, and about which creatures have it and which do not, and why. I must here for a moment give way to my friend Dr. Michael Shermer:

> Evolution also posits that *modern organisms should show a variety of structures from simple to complex, reflecting an evolutionary history rather than an instantaneous creation.* The human eye, for example, is the result of a long and complex pathway that goes back hundreds of millions of years. Initially a simple eyespot with a handful of light-sensitive cells that provided information to the organism about an important source of the light; it developed into a recessed eyespot, where a small surface indentation filled with light-sensitive cells provided additional data on the direction of light; then into a deep recession eyespot, where additional cells at greater depth provide more accurate information about the environment; then into a pinhole camera eye that is able to focus an image on the back of a deeply-recessed layer of light- sensitive cells; then into a pinhole lens eye that is able to focus the image; then into a complex eye found in such modern mammals as humans.

All the intermediate stages of this process have been located in other creatures, and sophisticated computer models have been developed which have tested the theory and shown that it actually "works." There is a further proof of the evolution of the eye, as Shermer points out. This is the ineptitude of its "design":

> The anatomy of the human eye, in fact, shows anything but "intelligence" in its design. It is built upside down and back wards, requiring photons of

> light to travel through the cornea, lens, aquaeous fluid, blood vessels, ganglion cells, amacrine cells, horizontal cells, and bipolar cells before they reach the light sensitive rods and cones that transduce the light signal into neutral impulses-which are then sent to the visual cortex at the back of the brain for processing into meaningful patterns. For optimal vision, why would an intelligent designer have built an eye upside down and backwards?

Whatever we think of the plausibility of Shermer's argument, it is clear that this kind of evolution is about enhancement not design. If it is one step at a time, then that is enhancement. It is this kind of enhancement thinking in combination with philosophical naturalism that makes evolution seem credible. Of course I do not know that the evolution from a simple eye to a complex eye can be explained as enhancement but Shermer is saying that sophisticated eye is merely an enhancement of a simple eye.

Although there are lots of intermediate steps in the design of the eye we have no idea about whether the large number of intermediate eyes will allow the eye to evolve. For all we know the extremely large number of intermediate eyes may only help us to imagine that the eye could evolve.

The shape of the eye includes simplicity, and no doubt there is an extremely large number of intermediate eyes that work, but there seems to be no way of getting a handle on such large numbers. Instead we can consider the design problem using the haemoglobin number. We know for example that performance complexity of the haemoglobin molecule must be less then prescriptive complexity if a cube, tetrahedron and many other kinds of distorted spheres work. Suppose that we have a billion other performance solutions. That means we are supposing that there are a billion other viable haemoglobin sequences and so we must divide the haemoglobin number by a billion. However, the haemoglobin number which is 10^{190} is so large that dividing by a billion (10^9), hardly makes a dent in the number. Although these numbers have been plucked out of the air they do suggest that there is potential for our imagination to get things wrong. What this means is that, even though there are lots of solutions we may need a large leap from one solution to the other regardless what our imagination tells us.

Suppose, instead, we work from the other direction and we say that we want a 50% chance of being able to move just one step (one amino acid change) towards improved or equal viability alternative solution then we would typically need access to say at least two other viable molecules one which would be more viable and one which would be less viable. For the haemoglobin molecule, this would mean that we need 2 in every 19*146 (or 1 in 1400) molecules to be viable. Here, the value 19 represents the fact that there are 19 alternative amino acids and 146 represents the number of amino acids in the chain. If we now assume that our current position is a random position in design space, then, when we make such an assumption, we reduce the complexity from the original 1 in 10^{190} to 1 in 1400 which would require 10^{187} viable intermediate states.

In other words, I can agree with Shermer that the human eye has an enormous amount of simplicity but I have no idea how much irreducible complexity remains. Unlike the first life form problem, there are sufficient intermediate states to allow us to imagine that there is an evolution solution, which is all that scientific thinking requires, because creationism is not science. In the above citation Shermer points out the design of the eye is upside down, but what Shermer means to say is that the design of the eye is probably upside down. He means given what he understands, the design of the eye is upside down. And so we must ask 'how probably' because when we know so little about anything, we are bound to find 'imperfection'.

Suppose that we wanted to change the software that drives a computer algorithm that plays chess. We can change the algorithm without destroying it, but we cannot change it step by step to achieve some other function unless the steps are sufficiently large. This is Behe's irreducible complexity and this is a feature of machines.

Behe's irreducible complexity seems like ordinary complexity which is nothing like evolution compatible complexity. Algorithms like the evolution algorithm may be able to find where a curve crosses the x-axis but they cannot play chess. This is because the evolution algorithm, like Newton's root-finding algorithm, is a feedback algorithm that generates a change in the required direction under certain conditions. If we are going to use a logical algorithm to solve this problem then we need to find the right direction with some sort of feedback algorithm. If the curve has a little complexity, say in the form of one or two peaks, or if it has a lot of complexity in the form of say 100 simultaneous curves, then we may need to go in the wrong direction and keep going in the wrong direction to get where we want to be.

Feedback may help to solve some simple enhancement problems but if the feedback points us in the wrong direction then we would do better to simply guess the answer with random trials.

The evolution algorithm will always take the best step from the current location, but in general it will not even take the best combined two steps from the current location. This is because the best combined two steps may require that the first step is downwards, and the best combined three steps may require some other combination, etc. Evolution only allows one combination from a particular starting position, which means the final destination achieved by evolution is dictated by the starting position. And the starting position is itself determined by chance.

Because evolution has no intelligence, it cannot climb down a small dip to go up a larger hill, if it comes to a dip or a rock it must stop. Evolutionists have long recognised these problems with the evolution algorithm, but they argue that there are ways around such problems. Evolutionists also agree that evolution cannot make irreducible complexity, but they argue that there are ways around irreducible complexity.

When Behe's book was first published, a number of examples appeared on the internet demonstrating how evolution of a mouse trap could evolve, because that is what the primary prejudice of science says that scientists must do. In earlier days, a favourite argument of the creationists was that a wing on a bird could not evolve. The idea is that small wings are useless or even a nuisance and therefore, there is no path from no wing to a wing. We have a dip in the path that says 'evolution cannot go any further' because evolution is an upwards climbing algorithm. The wing argument has been successfully refuted a number of times by evolutionists, by conjecturing credible scenarios of how the wing could have evolved. On p69-72 of his book, Behe cites the example of the rotating bacterial flagellum as an example of irreducible complexity. This structure apparently includes a rotor and stator to rotate the whip like flagellum and provide propulsion through a liquid. Spokesmen for evolution have counter-claimed that Behe's bacterial flagellum is not irreducibly complex because there are other non-rotating structures that are similar and therefore, a simple modification from one of these other structures could produce a functioning bacterial flagellum.

Even if it is true that no examples of complexity have ever been used to successfully argue that evolution cannot happen, complexity is still an impossible problem for the evolution algorithm to solve because all the solutions require that there is some path around the problem. In the end, we are only interested in whether naturalism can solve the problem, nevertheless, it is important that we draw a distinction between the evolution process and other natural scenarios, because one is a lawful creative algorithm that supposedly explains biological complexity and the other relies on chance happenings that can be invoked by our imagination to lend a hand to our scientific thinking. The process of evolution consists of the three link lawful process but a natural process can include the evolution process in combination with some other things that happen to occur at the particular time. In order to say that we cannot have a natural solution, we must somehow assert that we have a sufficiently difficult wing type problem or a sufficient number of difficult wing type problems or sufficient irreducible complexity, such that a natural solution is unlikely. The question of the size and number of such obstacles that will stop naturalism is then a question of time and chance.

We could suppose that we have another component that is similar to the component that we are trying to create. Or we could suppose that some non-irreducibly complex system was created that performs some task fairly poorly but that something changed which allowed the system to perform some other function, so that all the components became indispensable. These scenarios are ways around the problem of irreducible complexity.

A particular complement to the process of evolution may help to get around a particular problem, but complements that are not a lawful algorithm will not help with the solution of the design problem because any complement that is not a lawful algorithm is chance and chance destroys. Suppose we had a specific initial sequence of 146 amino acids and we wished to mutate this sequence into the haemoglobin prescriptive solution. If we take as our basic mutation the swapping of a single amino acid at any position in the sequence with an amino acid from a stockpile of all 20 amino acids, then we could form 19*146 of the possible $20^{146}= 10^{190}$ amino acid sequences in a single step. This would mean much more power for the evolutionary process, but if we are in a position of only one simple swap (of one amino acid) away from the target mutation the chances of getting the target state in the next step will be the same as it would be from the most remote state. Such scenarios may allow evolution to work in our imagination, but they cannot solve the problem of design of life if we believe that life is complex in the sense that we ordinarily use the word complex.

Evolutionists sometimes call on things like the Baldwin effect and other such complements to add power to the evolutionary algorithm, but again such complements are more about using chance and our imagination than solving the design problem. Apparently terms like synergistic epistasis, gene hitchhiking, truncation selection arise from an insufficiency of available time — some of these terms may be discussed in Walter Remine's Biotic Message. Although I have not read Remine's book, I do note that the assumed complexity in the Haldane dilemma calculations is only about 1:1000 rather than 10^{100} and increasing the complexity will make the problem of insufficient time even more severe.

Although there are a lot of genes that can be turned off by a mutation without causing problems, it makes sense to say that at the core design life is extremely complex. When I say 10^{100}, I have in mind such complexity and of course the design of such complexity cannot be achieved one step at a time. Evolutionists imagine that there is much evolutionary progress that can be achieved one step at a time because there is no other natural option. Evolutionists assume that complexities of the order of 1:1000 represent the complexity involved in the design of life because that is what enhancement requires.

The problem of the first life form is also a problem of ordinary irreducible complexity, except that we are not free to imagine that enhancement is the same as design. If we suppose that bacteria in our world are not the simplest self-sufficient life form, then we should be able to simplify it to a point where it is irreducibly complex and we should understand that it will still be an extremely complex machine. This means we are not free to imagine that a bacteria grew from nothing. The fact that we cannot solve this problem with our imagination is why Cairns-Smith in his book, *Seven Clues to the Origin of Life*, considers clay crystals as a precursor to biological life, because clay crystals can reproduce themselves. On page 39, Cairns-Smith explains that the problem he is trying to solve cannot be solved by chance, because the probability of a chance solution is extremely low, therefore it must be solved by the evolution algorithm:

> Even if amino acids (for example) had been in a 'probiotic soup', what use would they have been, long before their key use now (to make protein) had been hit on? *It is the whole machine that makes sense of its components.*
>
> **Point two**. Subsystems are highly INTERLOCKED within the universal system. For example, proteins are needed to make catalysts, yet catalysts are needed to make proteins. Nucleic acids are needed to make proteins, yet proteins are needed to make nucleic acids. Proteins and lipids are needed to make membranes, yet membranes are needed to provide protection for all the chemical processes going on in a cell. It goes on and on. The manufacturing procedures for key small molecules are highly interdependent: again and again *this* has to be made before *that* can be made — but that had to be there

> already. The whole is presupposed by all the parts. The interlocking is tight and critical. *At the centre everything depends on everything.*

If we do not already have life, then we cannot have evolution. Cairns-Smith argues that we do not really need life as we know it, but we need some way of implementing the mutation-selection-copy algorithm and to achieve this, he suggests that clay crystals may be the answer.

The problem of first life form seems absolute because we all know that there is no enhancement solution. Science gets to the fist life form by conjecturing infinite universes and clay crystals.

And from a single cell science gets to the many multi-cellular life forms by assuming an evolution compatible complexity that can be solved with a logical enhancement algorithm. In multi-cellular life forms, the problem of evolving may be more difficult than the evolution of the first life form, but because we know nothing about the details, scientific thinking easily solves such problems in our imagination. All we need is an enhancement algorithm and the assumption that life has a performance complexity of around 1:1000 and chance happenings to overcome the obvious obstacles that we can see.

There are reasons to suspect that multi-cellular life forms such as the human body may be much more complex than singled celled life forms. The human body is a complex machine, not just because of the human eye, brain and other structures, but because the human body can engender its like. The first cell divides into two and then each of these cells divides, etc, to eventually form a mass of cells. Though it should be impossible to believe, as the mass of cells grow, different organs, bones, joints, reticulation networks, skin cells, nerve cells, muscle cells, brain cells, etc, are formed.

Apparently, the early embryonic cells are polar with respect to the concentrations of chemicals carried in the cytoplasm and the cells divide so that the products of division are dissimilar — and this provides some direction for the necessary asymmetrically ordered development of an embryo from a single cell. Also, as cells become more specialised some genes are blocked by a protein covering so that only a fraction of the total number of genes in specialised cells are accessible for use by the cell. When the single cell develops into an adult human, the adult in turn can produce half-cells that combine with half-cells of the opposite sex to produce another embryo. Ecclesiasts 11:5:

> As thou knowest not what is the way of the spirit, nor how the bones do grow in the womb of her that is with child: even so thou knowest not the works of God who maketh all.

We get a glimpse what Solomon is talking about when we see a baby being born and what we see is nothing like a complexity of 1/1000. We should give ourselves a few moments to think on this problem. If we take away familiarity, then in no sense is it rhetorical to say that these things are much more fantastic than Joshua commanding the sun to stand still, or Jesus walking on water.

Science gets to the fist life form by conjecturing infinite universes and clay crystals and from there it gets to the many multi-cellular life forms by forgetting that we do not know 'how the bones do grow in the womb of her that is with child'. If man wanted to make a machine that did some of the things that an animal could do, then we would have a very complex machine. However if man wanted to add to such a machine the power to engender its like and maintain itself, then that would be an entirely different level of complexity. When spokesmen for evolution respond to Paley's argument by saying that biochemical machines are different from Paley's man made machines, this is the feature that spokesmen for evolution invoke.

Both Dawkins' analogy and the evolution algorithm can be represented by a hill climbing exercise. We can arrive at a very improbable location, even though we only take one very small step at a time. We know we can achieve a certain type of improbability (enhancement improbability) by the accumulation process, but Behe's question is really whether we should be using hill climbing to represent the complexity of biological life.

Dawkins argues that natural selection is the very opposite of chance, but natural selection selects - it does not create. Even though in some respect we can arrive at a very improbable solution via the hill climbing algorithm, we are not solving the design problem. Not if the design problem is say, making a mouse trap or a watch or a computer program or a radio, because these are all things that a mutation-selection-copy process cannot design.

Evolution-compatible complexity is represented by the peaks of smooth mountains in design space. Design complexity is represented by a typical computer program or a typical mechanical or electronic machine, or the mouse trap or the first life form. Instead of being like smooth hills, these examples are isolated in design space like the moon and planets are isolated from earth in space. We need a leap rather than a number of steps. Whether this sort of complexity is associated with some special feature of the design or if it is the basis of the fundamental function of the overall machine, we have the same problem.

'Design complexity', 'minimal complexity', 'performance complexity' and 'irreducible complexity' are similar terms. They are ordinary complexity numbers and all ordinary complexity numbers are anti-evolution numbers.

The above discussion concerns the details of science. The details are about determining whether or not evolution is the truth. Unlike the details mentioned in the previous chapter the issue of complexity is a very important issue. However it is also important that we understand the scientific approach to such details.

From the above it seems that science has the understanding that it must solve all problems that obstruct the theory of evolution and never find out what is wrong with the scientific solutions.

Shermer explains that a complex eye can evolve from a simple eye by insisting that the eye is an enhancement problem. That is, Shermer solves the problem by insisting that a 'complex' eye is no more complex than a simple eye. This is because the primary prejudice of science says that we can be confident that evolution happened.

Of course there are reasons to believe in naturalism, just as there are reasons to believe that 10 bricks plus 10 bricks equals 20 bricks. Given philosophical naturalism, it is reasonable to work backwards from the idea that there must be a natural solution. However, we know from history that naturalism can lead us astray. Philosophical naturalism is why science once believed in the magic of Darwinism. Philosophical naturalism is why the ancient Greeks believed that atoms could self-organise to create things. Philosophical naturalism is why both ancient Greek science and modern science believed in spontaneous generation. And a commitment to philosophical naturalism means that we will need extra effort to see what the arguments say without the help of philosophical naturalism.

Apparently there is more to this problem of truth than the primary prejudice of science. The fact that there are scientific theories like the infinite universe theory and the clay crystal theory, means that scientists are well aware of the fact that science is using the primary prejudice of science to solve the problem of creation and they are well aware of the fact that apart from the prejudice of philosophical naturalism the first life from does clearly say that there was a creator. The other issue here is apparently that many scientists are concerned about the 'subversion of science'.

It may be that somehow scientists are right not to agree with intelligent design theory. It may be that for whatever reason creationists arguments include a natural rhetorical advantage. Regardless of the rhetorical advantage of creationism, regardless of our confidence in philosophical naturalism, an ordinary understanding of truth means that philosophical naturalism is not acceptable as a hidden bias when we are investigating intelligent design because when we are investigating intelligent design we are investigating philosophical naturalism. Given that there is life on earth the faith of philosophical naturalism is bound to insist that evolution is the right place for all arguments to end up. Science needs to be able to say what the evidence seems to say.

Arguments are supposed to be about discovery. If we are going to allow the gravity of philosophical naturalism to answer all questions then why bother arguing? Why not simply assert that the fact that life exists proves that evolution happened?

It may be that somehow Behe is wrong, but if science had an ordinary understanding of truth it would be telling us that we would not believe that their is such a thing as evolution compatible complexity apart from the fact that life exists and that science believes in philosophical naturalism. Science should also be telling us that algorithms like the evolution algorithm cannot solve complex problems. Science should be telling us why adding complements to the evolution algorithm cannot help the theory of evolution solve the general design problem. Possibly science does not even understand such things because the job of science is to defend the theory of evolution.

Scientists, also put their faith in fellow scientists. When someone like Cairns-Smith assumes the evolution algorithm can explain the complexity of multi-cellular life forms given the first life form, then he may be putting faith in the judgements of his fellow scientists because that is a reasonable thing to do. And, of course, it is even easier to understand why people like Christopher Hitchens would put their faith in science.

I have heard Hitchens speak and read some of what he has written. When I heard about his death on the news it was not like hearing about so many other people that I hear about on the news, because I felt that I knew him. It was a shock to me, because it was my hope that he might one day get to change his mind, but instead he died placing his confidence in science. In my citation from his book, about a dozen paragraphs above, Hitchens says of the evolution process and the eye: "All the intermediate stages of this process have been located in other creatures, and sophisticated computer models have been developed which have tested the theory and shown that it actually 'works'." How does a scientists develop a computer model to test the mutation-selection-copy algorithm to create complexity and not discover that it cannot create ordinary complexity? How does the scientific community get to accept such statements? And who is going to take responsibility for Hitchens' destiny?

Although we must all agree that evolution cannot create ordinary complexity, philosophical naturalism is bound to insist that there is such a thing as evolution compatible complexity that creates designs. Philosophical naturalism is bound to assume that an enhancement algorithm can design otherwise the mutation-selection-copy algorithm would be entirely irrelevant to the design problem. Perhaps something like William A. Dembski's *Design Inference* argument will help to convince evolutionists that there is no such thing as evolution compatible complexity. I have not read Dembski's book at this point in time, but apparently his argument is not endorsed by mathematicians. The questions I would like to ask mathematicians who believe that a complexity of 1:1000 can explain life is: (1). Do you feel you do not have a firm understanding of the problem? If 1:1000 works then why not 1:2. (2). Does any of your thinking come from a trust in philosophical naturalism? (3) Does any of your thinking come from a trust in science? (4) What do you think your fellow mathematicians and scientists will think of you if you contribute to the subversion of science and perpetration of the idea that there is a supernatural creator? Would it be easier to think for yourself if you kept your opinion to yourself?

Chapter 6 WHAT IS SCIENCE?

Science is the discovery of empirical knowledge based on the principle of repeatability, and other reliable prejudices. The understanding that '10 bricks plus 10 bricks is equal to 20 bricks' relies on the logic of mathematics and the scientific prejudice that objects are conserved. The success of mobile phones, satellite navigation, air travel, televisions, computers, and the internet testify to the reliability of the prejudices of science.

We never see God intervening against scientific experiments. Naturalism prevails. If this was not the case then we would not be able to rely on any of the other prejudices of science including the understanding that 10 bricks plus 10 bricks equals 20 bricks. Reliability is a version of credibility and it is in the interest of reliability that we want to keep science free of the *influence* of religious texts and doctrines. We want science to be science not religion because we want science to have scientific reliability. If science has anti-religious implications then so be it. This is the way it must be with a few considerations and cautions.

As previously discussed one of the considerations we must keep in mind is that investigations are not assertions. If we are using philosophical naturalism to do our thinking then we need to make our audience aware of the fact that we are using philosophical naturalism to do our thinking. If we were considering a perpetual motion machine claim then good science allows us to use the laws of thermodynamics to pronounce an investigation a waste of time. However, if we are investigating the details of the perpetual motion machine then we are supposed to be investigating against what we know to be true.

Fundamentally, the problem of creation-versus-evolution is a problem of balancing 'the difficulties of creation by naturalism' against 'philosophical naturalism'. On one side, we have the difficulties of creation by naturalism and on the other side we have the understanding that naturalism prevails. If someone tosses a coin we say that there is about a 50% chance of the outcome being 'heads'. We are certain of this. If we repeat the toss 20 times and we get 20 heads in a row, we don't know what to think because we doubt the coin and when we doubt the coin we doubt everything. Evidence might show that someone was guilty of a crime, well beyond any reasonable doubt. If some evidence that the police have been framing the accused is brought to light, then some of us will think that there is a 50:50 chance that the accused is innocent and some may continue to think that there is 1 in a million chance that he is innocent, because these sorts of problems introduce wildly conflicting background probabilities.

6.1 Naturalism?

It is natural to assume naturalism, in our everyday life and in science. We all do that. But concerning the problem of creation, philosophical naturalism is confronted with the difficulty of the task of the natural creation of bio-chemical complexity? Creationists are stuck on the understanding that 'the existence of living things' is the beginning and the end of this argument. Scientists are stuck on the understanding that bio-chemical complexity is a special type of evolution compatible complexity and that there must be some way around the problem of the first life form.

As we have previously seen, Cairns-Smith is capable of providing a very open and honest perspective on the problem of the creation of the first life form, yet he is still confident that we will find a natural solution. On page 6 of *Seven Clues to the Origin of Life*, Cairns-Smith explains his confidence:

> It is a sterile stratagem to insert miracles to bridge the unknown. Soluble problems often seem to be baffling to begin with. Who would have thought a thousand years ago that the size of an atom or the age of the Earth would ever be discovered? Poor Dr. Watson was always being baffled by Sherlock Holmes' cases — as we all are by a good conjuring show. It is silly to say that because we cannot see a natural explanation for a phenomenon then we must look for a supernatural explanation. (It is usually silly anyway.) With so many past scientific puzzles now cleared up there have to be very clear reasons not to presume natural causes. Let us not say that the law was broken.

Cairns-Smith gets his confidence from philosophical naturalism. On page 234 of *Darwin's Black Box*, Behe writes:

A good example of disciplinary chauvinism can be seen in Robert Shapiro's fine book, *Origins: A Skeptic's Guide to the Creation of Life on Earth.* After presenting a very readable very devastating critique of scientific studies on the origin of life, Shapiro proclaims his steadfast loyalty — not to the goal of "explaining the physical world," but to science:

> Some future day may yet arrive when all reasonable chemical experiments run to discover a probable origin for life have failed unequivocally. Further, new geological evidence may indicate a sudden appearance of life on earth. Finally, we may have explored the universe and found no trace of life, or processes leading to life, elsewhere. In such a case, some scientists might choose to turn to religion for an answer. Others, however, myself included, would attempt to sort out the surviving less probable scientific explanations in the hope of selecting one that was still more likely than the remainder.

Shapiro and Cairns-Smith have an open faith in philosophical naturalism. Like Shapiro and Cairns-Smith, Dawkins and Dennett are also open about their faith in naturalism? Dawkins and Dennett (and perhaps Shapiro and Cairns-Smith) have a strong form of philosophical naturalism that discounts all religion factors.

Dawkins complains about the fuzzy thinking of 'religious scientists'. Religious scientists tend not to be so open about how they use philosophical naturalism in their thinking. A religious scientists may believe that there is some kind of god who many billions of years ago set the laws into operation and it may be that this provides the opportunity for religious scientists to define philosophical naturalism so that it is not a part of their thinking. Religious scientists also seem to have a poor understanding of the fact that 'science has a problem with a real God who can do things' because they have fuzzy way of thinking about the world.

Regardless whether we have a strong form of philosophical naturalism or a religious form of philosophical naturalism, if we place some credibility in things like Cairns-Smiths clay crystal theory, then we have a fully functioning form of philosophical naturalism. As previously explained the God-is-dead enlightenment taught us to think with philosophical naturalism because the God-is-dead enlightenment and philosophical naturalism are more or less the same things. Both were inspired from the success of the scientific revolution. In the preface of their *Science and Creationism* booklet NAS writes of such success:

> The tremendous success of science in explaining natural phenomena and fostering technological innovation arises from its focus on explanations that can be inferred from confirmable data. Scientists seek to relate one natural phenomenon to another and to recognize the causes and effects of phenomena. In this way, they have developed explanations for the changing of the seasons, the movements of the sun and stars, the structure of matter, the shaping of mountains and valleys, the changes in the positions of continents over time, the history of life on Earth, and many other natural occurrences. By the same means, scientists have also deciphered which substances in our environment are harmful to humans and which are not, developed cures for diseases, and generated the knowledge needed to produce innumerable labor-saving devices.

The 'by the same means' component of this argument can be interpreted as an argument for 'typical science' (or 'successful science') and the 'cause and effect' component of this argument can be interpreted as an argument for 'philosophical naturalism'. We could say that in the past, when we have seen that things are difficult to explain we conjectured a supernatural explanation, however, we are working our way through all these conjectures and dispensing with the need for supernatural explanations because our laws are repeatable and they are not subject to divine intervention.

Science has been extremely successful at providing explanations for natural systems that once were considered impossible to explain without supernatural intervention and therefore the creationist claim that "creation is a special problem" is ad hoc (or contrived). Based on the success of naturalism, scientific thinking understands that we should expect to be able to dispense with the need for a creator to explain creation, because we never see God intervening against repeatability.

Philosophical naturalism, like the idea that 10 bricks + 10 bricks = 20 bricks, is an inductive argument and therefore, it can be extremely probable but not mathematically certain. Inductive arguments can be extremely probable but in the case of philosophical naturalism, we do not achieve even a moderately probable status because of Hume's problem. The creationists' model, which asserts a creator (and not a operator-and-maintainer), effectively claims that 'operation-and-maintenance-law-systems' is the true fundamental variable, whereas 'natural-systems' is only a sensible quality/association, because as it happens, the majority of natural systems science has considered are to do with operation-and-maintenance laws. That is, creationism by definition claims that creation is a special problem because it is not a problem concerning 'operation-and-maintenance of the universe' and therefore, the success of science in understanding operation-and-maintenance laws has no relevance whatsoever to the truth of creationism. This is similar to the section 1.1 example of scientists in the southern hemisphere noting that certain fruit trees fruit at approximately the same time each year (say around December) and attempting to use their generalisation in the northern hemisphere.

Hume said that we cannot trust patterns, because we do not identify the secret (true) causes of patterns. Mill showed us how to determine the secret (true) causes of empirical patterns. And as it turns out, there is such a thing as type 1 induction that works even when we do not have an understanding of the true secret causes. However, type 1 induction fails when we notice that there is something critically different about the next instance. And we should all be able to understand that the creation instance is not like the countless numbers of operation-and-maintenance instances.

The same NAS citation appears towards the end of the introduction where I supported it with philosophical naturalism. Because I am writing to a secular audience, I have used a subtle form of philosophical naturalism in many of my arguments to express what scientists believe.

The introduction to Chapter 4 initially defines philosophical naturalism as the idea that naturalism is probably all there is, but that definition is itself based on inappropriate inductive thinking because it does not see the difference between creation and operation-and-maintenance.

Although my philosophical naturalism arguments are defective, I have tried to keep my philosophical naturalism subtle just NAS keeps their philosophical naturalism subtle. Cairns-Smith and Shapiro do not keep their philosophical naturalism subtle because they are open about their philosophical naturalism. From page 250 of *Darwin's Dangerous Idea* Dennett writes of one of Dawkins scientific assumptions:

> Does Dawkins know this? Does he know that the postulated intermediates were less fit? Not because he has seen any data drawn from the fossil record. This is purely a theory-driven explanation, argued a priori from the assumption that natural selection tells us the true story some true story or other — about every

> curious feature of the biosphere. Is this objectionable? It does "beg the question" but what a question it begs! It assumes that Darwinism is basically on the right track. (Is it objectionable when meteorologists say, begging the question against supernatural forces, that there must be a purely physical explanation for the birth of hurricanes, even if many of the details so far elude them?)

That is not subtle. Dennett boldly declares that his confidence is based on a failure to distinguish creation from operation-and-maintenance. Instead of understanding that hurricanes belong with operation-and-maintenance class and creation belongs in the creation class, Dennett suggests that they both belong in the 'difficult to explain class'.

We need to adjust our definition of 'philosophical naturalism' so that it means inappropriate naturalism. With this new understanding we can say that philosophical naturalism is a prejudice that gets its confidence from a failure to distinguish creation issues from operation-and-maintenance issues. In other words science makes creation a part of operation-and-maintenance, because science insists that 'creationism is not science'.

Except for philosophical naturalism it does not make sense to imagine that there is a natural solution to the problem of creation. Except for philosophical naturalism it makes no sense to conjecture that clay crystals may have been involved in the creation of the first life form. Except for philosophical naturalism it makes no sense to assume that a trial-and-error algorithm can design multi-cellular animal life forms that can reproduce.

We can allow that philosophical naturalism is a valid religious theory because there must be some chance (where chance is associated with some limited information we have about the theory and not the theory itself) that the theory is true. It is possible that when scientists in the southern hemisphere noted that certain fruit trees fruit at approximately the same time each year (say around December) that their generalisation would continue to work for the northern hemisphere. However, confidence cannot be provided by the pattern, and it is clear that Dennett and other scientists get their confidence from the pattern. It is clear that Dennett expects the pattern to inspire the same confidence in his readers.

Instead of noticing that creation seems like it might be both a difficult problem and a different kind of problem and allowing that there might be a God who might want laws to control the atoms and photons and the planets and the stars, philosophical naturalism starts from the assumption that there is no God who can do things. Philosophical naturalism insists that, because there are natural laws that determine how the universe runs, we should therefore believe that natural laws must also determine how the universe was made. And of course if we do not distinguish operation-and-maintenance from creation, then we will have no choice but to interpret the pattern of naturalism inappropriately. If we do not distinguish creation from operation-and-maintenance then we are saying we have no reason to suspect that the creation instance is a special instance.

This should be a problem for philosophers. On page 55 of his *Philosophy of Biology*, Elliott Sober (the Hans Reichenbach Professor of Philosophy at the University of Wisconsin, Madison) argues:

> When creationists use the current incompleteness of evolutionary theory to argue their position, they should be asked this question: Why do you think that no scientific explanation will ever be developed for the phenomenon in question? Our current ignorance is no evidence of the truth of any explanation, creationist or otherwise. The fact that we currently do not understand various facts

> about life is no reason to think that God has intervened in life's history. At the same time, the past successes of scientific explanation [of maintenance laws] suggest that what now is inexplicable [namely the problem of creation] may eventually be brought within the scope of scientific understanding. Significant advances already have been achieved in understanding the origin of life, the evolution of sex and the process of ontogeny. Many questions remain to be answered. Can science do the job? It is not utterly crazy to think it can. The track record of science offers grounds for optimism; nothing of the kind can be said concerning the track record of "creation science."

We can allow that the poor track record of the arguments of creationists allows us to make some prejudgements about the arguments of creationists and yes, in some sense, that would allow us to make a very weak prejudgement about creationism. However Sober either believes that the idea that 'God has made operation-and-maintenance laws so that he does not need to supervise every photon and electron in his universe' is an incredibly ad hoc idea, or he cannot see that the creation problem is different to all the other operation-and-maintenance problems that have been explained. Sober thinks that if God created then we should have a different track record for science to explain all the operation-and-maintenance problems. If we can remove ourselves from our philosophical naturalism for a brief moment then we will be able to suppose that God did design all life forms and that God created laws to operate-and-maintain this universe. It should be the case that philosophers would be of some help in addressing the problem of philosophical naturalism, except that cultural pressures dictate that evolution is a 'right place' for arguments to end up and spiritual pressures dictate that philosophical naturalism is the truth.

Creation is different to operation-and-maintenance. It is reasonable to assume that God would have operation-and-maintenance laws so that he does not need to supervise every photon and electron in his universe. And it is reasonable to assume that creation might require a creator, but unreasonable to assume that operation-and-maintenance laws can replace a creator.

Freethinkers use their philosophical naturalism to rule out God, which confirms their philosophical naturalism. In the context of the creation-evolution argument, the prior probability that there is a God, is the probability that there is a God assuming we do not need a creator to explain creation. And the prior probability of philosophical naturalism is the prior probability that naturalism can make walking talking self-reproducing self-maintaining biochemical machines assuming such machines were made by God or some other intelligent designer. Unless we want to use philosophical naturalism to enforce philosophical naturalism, there is no escape from using some word like 'God' because these arguments are about prior probabilities. In fact they should not be about prior probabilities because we should know better. However given an unreasonable faith in philosophical naturalism they should be about prior probabilities.

Philosophical naturalism is a prejudice. Prejudices are good because prejudices provide our confidence, except that, when the next instance happens to be the creation instance, then our prejudice is invalid. This is because our confidence in naturalism comes from an experience with operation-and-maintenance instances and the creation instance is critically different to operation-and-maintenance instances. I am not saying that creation is critically different because of extreme bio-chemical complexity but that it is different because it is creation. We can trust naturalism to explain how the universe is operated-and-maintained because we learn that from experience. When we assume we can trust naturalism to explain creation, then we are effectively saying that because our experience teaches us that God has laws of the universe to operate-and-maintain the universe (so that he does not need to be occupied with supervising every photon and electron in this universe), we can be sure that God also has laws to explain how things were created. Because apples always fall to the ground under natural law, and because we are familiar with the fact that apple trees can grow, scientific thinking is confident that apple trees evolved under natural law.

Kant and Laplace conjectured natural laws to explain the creation of the planets. Kant and Laplace may have had some understanding of the difference between creation and maintenance, but it is clear that our modern culture does not. Some of us have the kind of philosophical naturalism that resists the idea of a real God who can do things, and some of us have a stronger form of philosophical naturalism that resists both the idea of a real God and the idea of a fuzzy do-nothing god. Both kinds of philosophical naturalism do not recognise the difference between creation and maintenance. Both forms of philosophical naturalism have the understanding that 'creationism is not science'.

On page 28 of his book *The God Delusion,* Dawkins writes that: "The dictionary supplied with Microsoft Word defines a delusion as 'a persistent false belief held in the face of strong contradictory evidence, especially as a symptom of psychiatric disorder'. The first part captures religious faith perfectly. As to whether it is a symptom of a psychiatric disorder, I am inclined to follow Robert M. Pirsig, author of *Zen and the Art of Motorcycle Maintenance*: 'When one person suffers from a delusion, it is called insanity. When many people suffer from a delusion it is called Religion:'" But in fact Dawkins' philosophical naturalism, as the essence of the God-is-dead enlightenment, is a delusion that all westerners to some degree share. Our faith in naturalism was only ignorantly-probable reasoning (assuming that for whatever reason we do not see anything different about the creation instance), but it needed to be sufficiently probable to be able to stand up against the problem of the natural creation of the first life form and the complexity of advanced life forms. That is, even if the first life form was much less complex, we would have no reason to presume naturalism, because our enormous experience that tells us that naturalism prevails is not relevant to the problem of creation. The millions of examples of how naturalism prevails do not count when it comes to the problem of creation, just as millions of observations of fruit trees fruiting in December in our Southern hemisphere do not count when we want our prediction to apply to the Northern hemisphere. In these sorts of situations, we might say that the inductive inference is superseded by new information, except in the case of the problem of creation, the inductive inference has always been superseded and therefore, it makes more sense to say that it is a delusion.

Philosophical naturalism claims that because we do not need a God to explain how things are maintained; we do not need a God to explain how things were created as if the event of creation was like the law of gravity. Modern science understands that because we do not need to invoke a God to explain how the universe is operated-and-maintained, we do not need to invoke a God to explain how the universe was created.

We use the word *creation* because life does not seem like operation-and-maintenance laws in action. Life seems like it is something completely different from gravity for example. Science cannot solve the problem of the first life form but philosophical naturalism can. Science cannot solve the problem of creation of ordinary complexity but philosophical naturalism can. Philosophical naturalism is apparently more persuasive than logic because even western mathematicians do not seem to notice mathematical absurdities created by philosophical naturalism.

Dawkins and Dennett (and perhaps Shapiro and Cairns-Smith) have a strong form of philosophical naturalism unlike many religious scientists. People like Dawkins and Dennett are sometimes aware that they get their confidence from philosophical naturalism because Dawkins and Dennett are open about where their confidence comes from, whereas religious scientists are not always so quick to understand that philosophical naturalism is a part of their science.

Professor Phillip E. Johnson, in his book *Darwin on Trial,* has already argued that philosophical naturalism is in essence the philosophy of modern science and he claims that the idea that 'evolution must occur by naturalism' comes directly from this philosophy. However, instead of opening up more space for creationists to dig their heels in, science saw the danger and came up with a scientific response. In the her article '*DARWIN ON TRIAL: A Review*' Eugene C Scott, executive director of the *National Centre for Science Education* responds to Johnson's accusation of philosophical naturalism:

> Although science has made great progress by limiting itself to explaining through natural causes, Johnson would have us allow the occasional supernatural intervention for those phenomena that causes problems for his particular theology. Though he has no objection to natural explanations for how fluids pass through a cell membrane, we need to leave room for a "whimsical creator"(p.31) or God's "inscrutable purpose" (p.71) to explain the origin of life, or the origin of complex body plans. Confusing unsolved with unsolvable, he wishes us to stop seeking natural causes for these phenomena. Johnson wishes to take us back to an earlier, less scientifically productive

> time when science had to wait for religious approval before daring to explain something that was religiously sensitive.

The above argument is a form of philosophical naturalism, even though Scott is addressing Johnson's accusation of philosophical naturalism. This is because the essential feature of philosophical naturalism is that it does not see the difference between things like fluids passing through membranes, and the problem of creation. Like Dennett's argument cited above, this argument puts unsolved maintenance problems and unsolved problems of creation into a common unsolved class. This kind of scientific confidence comes from not noticing when there are good reasons why we should not extrapolate the pattern to new circumstances.

The above article comes from a very old NCSE document and I have not been able to find the article on the current NCSE website. This may mean that Scott no longer agrees with her argument, but my point is that philosophical naturalism infiltrates our thinking and gives us confidence without us being aware of where our confidence comes from.

We do not need to invoke philosophical naturalism to solve problems, because philosophical naturalism invokes itself because philosophical naturalism is about background probabilities, confidence, and a spiritual error. Because Scott does not distinguish maintenance from creation, she has no choice about how the background probabilities inappropriately work on her thinking. This is to say that because she can be confident that science will be able to determine a natural explanation for how fluids can pass through a cell membrane, she can be confident that science will be able to determine how life was created. That kind of thinking seems valid to Scott because she has already ruled out the God solution.

Despite Johnson's attempt it seems that many scientists have not been able to find any philosophical naturalism in their science? This is because science is insensitive to wrong arguments that end up in the right place. The current NCSE response to the philosophical naturalism accusation, which is apparently endorsed by NAS, has been to insist that science uses 'methodological naturalism' rather than 'philosophical naturalism'. The following arguments have been taken from the NCSE articles titled *Darwin Prosecuted: Review of Johnson's Darwin on Trial* and *Keep Science Free from Creationism*:

> [Darwin Prosecuted: Review of Johnson's Darwin on Trial] Johnson confuses the necessary methodological materialism (or naturalism) of science with philosophical materialism/naturalism. Science neither denies nor opposes the supernatural, but ignores the supernatural for methodological reasons. The history of science has shown that progress comes from logical and empirical study rather than reference to revelation or to inner psychological states. That's how we play our game; his basketball won't work on our baseball field. The essence of science is empiricism and control of variables, and if there is an omnipotent God, it certainly can't be controlled like temperature or humidity. Science has made a little deal with itself: because you can't put God in a test tube (or keep it out of one), science acts as if the supernatural did not exist. This methodological materialism is the cornerstone of modern science.
>
> Materialism also is the cornerstone of mathematics. It may be the case that God caused 2+2 to equal 4, but ultimate cause is irrelevant to the mathematical applications that can be made to 2+2 = 4. Scientists no more accept supernatural explanations for phenomena than mathematicians would accept a solution to a

mathematical problem based on revelation. Yet no one claims mathematics is antireligious. Neither is science.

Johnson fails to recognize the necessity for methodological materialism, because of his concern for philosophical materialism's attack on his theology. The process of evolutionary change, like any scientific process, must be studied without reference to the supernatural. Johnson is certainly welcome to criticize philosophical materialism if he wishes to, but such a criticism is irrelevant to science.

This conflation of methodological materialism with philosophical materialism also confuses two very different types of causation: proximate and ultimate. Science deals only with proximate cause; religion deals with ultimate cause. If God produced the Big Bang, or the first replicating molecule, science can tell us nothing about this, because the supernatural is outside of the realm of things that science can explain. Statements of ultimate cause may be true or false or in between, but they are not testable through the method of scientific discovery. They must be tested using other canons of thought, if they are testable at all. Johnson states that it is possible that God created through evolution, an alternative to a "fully naturalistic evolutionary process," but he doubts it.

This, however, is the position embraced by the majority of main-line Protestant and Catholic theologians: theistic evolution. I define theistic evolution as evolution governed by natural processes but begun by and/or guided by God. (There are many varieties of theistic evolution, as there are many nuances of the understanding of God in Christianity.) We can understand the natural processes using the methods of science, but we cannot understand the ultimate cause.

[Keep Science Free from Creationism] Resorting to the supernatural violates a major canon of modern science: explain only through natural causes. The reason is not anti-religious, but purely practical: better answers are found when only natural causes are specified.

Consider: if I grow two plots of corn, fertilize only one, and find that both yield the same number of ears, how do I explain my results? I can examine the chemical content of the fertilizer, and find that it contained no nitrogen, or I can say, "God wanted both plots to produce the same number of ears." Well, maybe so.

So I plant two more plots, fertilize only one, and this time the fertilized plot produces more ears. How do I explain this? Looking for natural causes, I might find that this batch of fertilizer has nitrogen, and maybe I can make a generalization to test further. But if I allow supernatural explanation, I have to consider that maybe God did it. Where would this get me? How can I establish general explanatory principles, like corn needs nitrogen to grow well, if I can explain away my results by invoking a capricious Creator? If I am to understand the natural world, I have to conduct my science as if only natural forces affected my subject.

And indeed, the world appears to operate according to regularities — it looks like God doesn't reach down and arbitrarily mess up experiments. So we don't need to look for miracles: just keep trying to find the natural explanation. In Pandas, Meyer claims that scientists don't allow supernatural explanation because the supernatural is not observable. Nonsense. Particle physicists study phenomena they can't directly observe, and so do many other scientists. But you can't (scientifically) study variables you can't test, directly or indirectly. You can't use supernatural explanation because you can't put an

> omnipotent deity in a test tube (or keep it out of one.) As soon as creationists invent a "theo - meter" maybe then we can test for miraculous intervention.

There are two background issues that should be clarified before looking at these arguments:

- These arguments use a concept of philosophical naturalism that excludes the super-natural from physics and chemistry, whereas the essentials of philosophical naturalism is actually the unjustified belief that anything other than the work of natural laws is *always* highly improbable. The supernatural is highly improbable but it is not always highly improbable. Philosophical naturalism is about not noticing the different circumstances. Using a definition that sorts out the confusion says that even Scott's definition of philosophical naturalism is philosophical naturalism, because her definition does not see the difference between creation and operation-and-maintenance.
- It may be correct to say that all scientists use methodological naturalism, but what we are really interested in is whether people get scientific confidence from philosophical naturalism. We know, for example, that there are scientists who are open about their philosophical naturalism and we know that Scott uses philosophical naturalism in her thinking even though she denies the influence of philosophical naturalism on science.

If we were to say that God designed Newton's law of gravity, this does not help science, because science wants explanations to be useful. We, therefore, use natural explanations because they are useful, because with natural explanations we can extend our experience to make predictions about other situations (using the principle of repeatability).

The explanation for gravity in terms of Newton's laws (or ears of corn in terms of nitrogen), does not contradict the idea that 'God might have made the law'. We could say that 'God designed the planets to move in accordance with the inverse squared law of ...' and that 'God designed the plants to respond to nitrogen' without introducing a contradiction, but of course we do not do that because we do not gain anything by attaching God to such statements.

The real problem with the statement that 'God wanted both plots to produce the same number of ears' is that science is interested in the truth and we will never have any evidence that such God statements are true, not because God is attached to the statement but because of how the statement is constructed. We could construct similar scientific statements simply by saying that some special natural effect travelling about the universe happened to come in contact with the corn.

Of course concerning issues of operation-and-maintenance such as growing corn, appropriate naturalism requires that we do not consider whether or not God has intervened. However, if we say that God made all life forms (not god evolved all life forms through a process of divinely assisted evolution), then we have an immediate contradiction with naturalism. Instead of two parallel explanations, one which tries to identify ultimate causes (God) and one that tries to identify useful causal criteria, we have two contradictory explanations. Suppose that God created, how do we then make sense of the idea that 'science cannot identify God because God is an ultimate cause'? Again, it seems that we need philosophical naturalism and a determination to end up at the right place to explain how science could be confused about such simple issues.

According to another NCSE article *My Favourite Pseudo-science*, "Science is nothing if not practical. The explanations that are retained are those that work best, and the explanations that work best are ones based on material causes. Nonmaterial causes are disallowed....Another reason for restricting ourselves to natural explanations is practical. It works. We have gone a long way towards building more complete and, we think better, explanations through methodological naturalism, and most of us feel that if it ain't broke, don't fix it". Again, Scott does not recognise the difference between creation and maintenance, and she does this in an article about how science is not based on philosophical naturalism. Furthermore, the claim that scientists use natural explanations because natural explanations work is classical 'inductive thinking'. Inductive thinking is fine except that in this case, we should say that naturalism works if we stick to operation-and-maintenance issues. Scott sees the problem of creation as being like all the other problems of science, except that it has not been solved yet, which is more philosophical naturalism. Another NCSE document *Facts, Faith, and Fairness*, which is no longer available at the NCSE site, but is available at the *SkepticReport* site, says:

> Scientists who investigate the past must proceed in the same way that detectives work when solving crimes without witnesses. In such cases, detectives must assume that no supernatural forces were involved. Suppose you are accused of a murder and you have overwhelming evidence to prove that you were 3,000 miles away from the scene of the crime when the murder was committed. But the prosecutor ignores this rational evidence, and claims that you made yourself invisible, flew at the speed of light to commit the murder 3,000 miles away, and returned an instant later. How could you defend yourself? Could you prove that you did not have these powers? No – it is impossible to prove or disprove something outside

> the realm of rational investigation. Any judge who heard a prosecutor accuse a defendant of using supernatural powers to commit a crime, would immediately rule that the accusation is inadmissible in court. In just the same way, the explanation of material facts by supernatural forces is not admissible in science. Religious explanations of the universe, in contrast to science, are based upon belief in certain forces that are beyond the realm of human understanding. Many religions also depend on a faith that certain documents are infallible.

We can all agree that the prosecutor who ignores the rational evidence, and claims that 'the accused made himself invisible, flew at the speed of light to commit a murder 3,000 miles away, and returned an instant later' has dropped too many prejudices but if we want ordinary truth rather than scientific truth, then we must agree that the prosecutor's ludicrous thinking is valid thinking when it comes to the problem of creation. The problem of philosophical naturalism is not that we use such probable thinking but that we use such probable thinking without recognising that the creation problem is fundamentally different to the operation-and-maintenance problems that we are so familiar with, which is EXACTLY what this *Facts, Faith and Fairness* article does. This kind of thinking comes from deeply entrenched philosophical naturalism.

Scott uses the concept of methodological naturalism to hide philosophical naturalism from herself and her audience, when the difficulty with philosophical naturalism is that it is mostly a hidden part of our thinking. Philosophical naturalism is more like a gravity than a deliberated bias. A brick layer does not consciously invoke the understanding that 'objects are conserved' when he wants to know how many bricks he has left because that is a hidden part of his thinking. Philosophical naturalism provides inescapable background probabilities that effect credibility judgements. Philosophical naturalism has a powerful influence on Scott's thinking even when she is in the middle of denying the influence of philosophical naturalism.

It is difficult to investigate if there are Gremlins in the attic because we all know that there are no Gremlins in the attic. And it will be almost impossible to investigate if there are Gremlins in the attic if the investigators can convince themselves that their thinking is not influenced by the prejudice that there are no Gremlins in the attic.

There should be no need for the above examples because obviously philosophical naturalism is the reason why science has some respect for the idea of infinite universes and clay crystal theories (to create the first life form). It is absurd that science would have any respect for arguments that deny the influence of philosophical naturalism on science. The explanation for the acceptance of this absurdity is apparently that science is insensitive to wrong arguments that end up at the right place.

When creationists point out the problems or short falls in the evolutionary arguments then spokesmen for evolution insist that creationist arguments are 'arguments from ignorance' or 'God of the gaps' arguments when the God of the gap's response is the essence of philosophical naturalism. For sure, we can say that if people use God to fill in some gap in problem of operation-and-maintenance of the universe, we can respond by saying that 'naturalism tells us that God does not maintain things'. That might not always be the case but it is an appropriate prejudice. However, when naturalism argues that there is no such thing as unsolvable, and 'unsolved' always translates as 'currently unsolved', even when the specific issue being discussed, concerns the problem of creation, that is inappropriate naturalism. The NCSE article *Science and Religion, Christian Scholarship and Theistic Science* trips over all the issues again in an article that argues that philosophical naturalism does not influence science:

> Advocates of theistic science would like to change all this by allowing "God did it" as a scientific, not merely theological, statement. They do stipulate, however, that God's hand not be invoked capriciously. Plantinga and others suggest that most of the time God operates using secondary causes, but room must be left for the occasional miracle. It is not coincidence that these allowances for miraculous interventions seem to focus around the topic of evolution: Moreland points out, for example, that "theologians have little interest in whether a methane molecule has three or four hydrogen atoms" but because God "designed the world for a purpose", he "has directly intervened in the course of its development at various points ([for example], in directly creating the universe, first life, the basic kinds of life, and humans)" (for details, readers should consult Moreland, *Creation Research Journal,* 1993...)
>
> Echoing this approach, Meyer separates science into two kinds: "historical" and "operational" ("empirical"). Operational science is the familiar everyday science exploring the processes and mechanisms of how the universe works, and miracles are not expected to be discovered. Both theists and nonbelievers would conduct operational science in the same fashion. Historical science, on the other hand, deals with nonrepeating events such as speciation events in the fossil record, the explosion of the Pinatubo volcano, the appearance of a supernova and so forth. Of course historical sciences can be studied scientifically: there may have been only one observed eruption of Pinatubo, but there certainly can be a science of volcanic eruption that can be used *to explain*

Pinatubo. Similarly, only once in history did a population give rise to genus *Equus*, but we can still derive theories from this and similar events to explain macroevolution.

More for theological than scientific reasons, God's direct hand is allowed in historical science, though strongly discouraged in operational science. This is so, it seems to me, because it is primarily historical sciences (like evolution) that have serious consequences for certain conservative Christian theologies.

When do theistic science proponents invoke a miracle? When they can't figure out a materialistic explanation. As Plantinga states,

> "Why couldn't a scientist think as follows? God has created the world, and of course He created everything in it directly or indirectly. After a great deal of study, we can't see how he created some phenomenon P (life, for example) indirectly; thus probably he has created it directly." (Plantinga, 1997)

Plantinga's position is given in more detail in. "Methodological Naturalism? Part 2", *Origins and Design*, 1997; 18(2):34 (footnote 63).

It is telling that three of the topics most frequently cited as requiring direct divine intervention (and cited by Moreland, above) are ones for which there is not yet consensus on a fully naturalistic explanation: the Big Bang, the origin of life, and the Cambrian explosion. The fourth topic is the origin of humans, upon which there actually is quite a good consensus, though the degree of agreement is not well-known to the general public.

It is fair to state that "theistic science" is a form of special creationism, or even creation science that cleverly ignores issues divisive to anti-evolutionists - such as the age

> of the earth - and focuses on unifying issues - such as the importance of God's hand in the universe. Theistic science has not been uniformly embraced even within the ranks of conservative Christians. (For critical reviews, see Howard Van Till, "Special creationism in design clothing: A Response to *The Creation Hypothesis*", Perspectives on Science and Christian Faith, 42:p.123-131, June, 1995; and D.F. Siemens, Jr., "On Moreland: Spurious freedom, mangled science, muddled philosophy", *Perspectives on Science and Christian Faith*, 49:pp 196-199.)

These 'God of the gaps' responses to creationist arguments convey the understanding that 'evolution is typical science' and they convey the understanding that 'the idea of a creator' is inspired by religion rather than because we need a creator to explain creation. God of the gaps responses are inspired by philosophical naturalism and they construct an unbeatable naturalism. An unbeatable naturalism is logical, given philosophical naturalism, because philosophical naturalism tells us that what seems to be an incredible explanation should be promoted to a position of certainty if it is the only possible natural explanation. Although unbeatable naturalism puts a stop to all progress in the God direction, spokesmen for evolution use the 'God of the gaps' objection in the name of scientific progress. According to the previously cited NCSE *My Favourite Pseudo-science* article:

> Also, being able to say, "God (directly) did it" is a "science stopper", in the words of philosopher Alvin Plantinga (2001). To say "God did it" means one does not need to look further for a natural explanation. For example, creationist literature abounds with criticisms of origin-of-life research. Because scientists have not yet reached a consensus on how the first replicating molecule came about, creationists argue, this is an intractable problem that should just be attributed to "God did it". Well, if we

> stop looking for a natural explanation for the origin of life, surely we will never find it. So even if we have not found it yet, we must nonetheless slog on....Push a creationist argument far enough, and sure enough, it will become necessary to resort to a miracle. But miracle-mongering cannot be part of science.

More than 150 years have passed since Darwin and we are still going around in circles. If we truly want progress, rather than scientific progress then science must understand that although miracle mongering cannot be a part of science that is concerned with the operation-and-maintenance of the universe, miracle mongering must be allowed to explain creation. If we understand that we should have in mind credibility and ordinary truth rather than scientific truth, then we should understand that what is a very reasonable basis for science namely that 'we should only expect a supernatural solution if a natural solution is possible' is delusional thinking for a creation discovery programme.

Instead of understanding that 'naturalism' is an entirely unreasonable assumption for creation, spokesmen for evolution tell us that we cannot say that God did it even if it seems like he did do it, but not because the world is going to end or the universe is going to fold in on itself. Instead, spokesmen for evolution say that we cannot postulate the supernatural because 'we will stop looking for natural causes'? It is true that if the Gremlins are undetectable, then elimination of the Gremlins by finding an alternative explanation may be the only option for progress, but if undetectable Gremlins are the cause, then the only option for progress will be progress in the non-science direction? If science absolutely insisted that God created the planets and somehow made it mandatory for all scientists to think that way, then science would stop there. Something like that was attempted when the western world was ruled by the Catholic Church. It seems that cultural and spiritual forces are far more effective than the rule of law because in the modern world almost all scientists obey the 'creationism is not science' mandate.

If God did create the planets, then science has already been stopped. What matters then, is whether science has been appropriately or inappropriately stopped. On page 159-160 of *The God Delusion* Dawkins writes:

> Such work would never be done if scientists were satisfied with a lazy default such as 'intelligent design theory' would encourage. Here is the message that an imaginary 'intelligent design theorist' might broadcast to scientists: 'If you don't understand how something works, never mind: just give up and say God did it. You don't know how the nerve impulse works? Good! You don't understand how memories are laid 'down in the brain? Excellent! Is photosynthesis a bafflingly complex process? Wonderful! Please don't go to work on the problem, just give up, and appeal to God. Dear scientist, don't *work* on your mysteries. Bring us your mysteries, for we can use them. Don't squander precious ignorance by researching it away. We need those glorious gaps as a last refuge for God.' ...After listening to Behe, Rothschild eloquently summed up what every honest person in that court room must have felt:
>
>> Thankfully, there are scientists who do search for answers to the question of the origin of the immune system ... It's our defense against debilitating and fatal diseases. The scientists who wrote those books and articles toil in obscurity, without book royalties or speaking engagements. Their efforts help us combat and cure serious medical conditions. By contrast, Professor Behe and the entire intelligent design movement are doing nothing to advance scientific or medical knowledge and are telling future generations of scientists, don't bother.

Before getting back to the topic of stopping science, it must be noted that amongst his philosophical naturalism, Dawkins makes a valid point. Even though we should be assuming irreducible complexity is everywhere in design, any examples of irreducible complexity must stand on their own, independent of a belief in irreducible complexity. Otherwise such examples become a kind of backwards thinking designed to take creationists arguments to the right place.

Although Dawkins is open about his use of philosophical naturalism in his arguments, he still needs to be able to distinguish creation from operation-and-maintenance if he is going to criticise creationism. This is to say that even though Dawkins is open about his philosophical naturalism, he needs to be able to recognise that while he may have his own religious reasons for asserting philosophical naturalism, there are no logical reasons to assert philosophical naturalism. Furthermore, if philosophical naturalism was a true faith rather than a delusion, then Dawkins (and Rothschild) would be able to see the difference between creation and maintenance, and they would not be confusing the investigation of how things are made with the investigation of how things work.

Science has been inappropriately stopped because ordinary progress should be about ordinary truth rather than spanning all gaps with philosophical naturalism. The concept of 'evolution compatible complexity' is a gap, but despite attempts at modelling evolution with computer programs, we cannot even be sure that science is aware that there is such a gap. The problem of the first life form is another gap but that is not a problem to philosophical naturalism because infinite universe theories or clay crystals can span that gap.

Ordinary progress allows that we can say that God did not do it but only if it seems as if God did not do it. Scientific progress says that, if the evidence speaks against religion, then so be it, but when the evidence suggests a 'creator' then, 'creationism is not science'.

If our spokesman for evolution could put their philosophical naturalism aside for a moment then they would be able to understand that the idea that clay crystals evolved into a bacteria is an utter absurdity. Philosophical naturalism provides credibility, except that there is no good reason to believe in philosophical naturalism.

Science has been inappropriately stopped because the pattern of naturalism leads us to consider naturalism as probable and supernaturalism as improbable, which creates background probabilities. The *NCSE List of Voices for Evolution,* article supports the understanding that philosophical naturalism is not just the delusion of western science, but it is the delusion of western society including various religious organisations. The *Theopedia* website (a Biblical Christianity site) says of the God-of-the-gaps argument:

> God of the Gaps arguments are a discredited and outmoded approach to apologetics, in which a gap in scientific knowledge is used as evidence for the existence of God.
>
> Before the scientific revolution of the last four centuries, such arguments were commonplace and widely accepted, presumably because the "gaps" were large and showing no signs of shrinking. A lightning bolt crashes down, the peasants working in the field cross themselves and say "well, we don't understand that, so it must be God."
>
> From a philosophical point of view, the inherent problem with a God of the Gaps apologetic is that it relegates God to only a portion of creation, the portion that we don't understand yet. It places the apologist at a disadvantage by ignoring how the underlying patterns in the things we understand speak to the work of the Creator God. It also

> denies, in effect, the Christian view of science, which is that science is "thinking God's thoughts after him"; it does this by suggesting that we can only see God in the areas of nature which we do not understand, rather than seeing him most clearly in those which we do understand.
>
> From a pragmatic point of view, the main problem with a God of the Gaps apologetic is that the gaps are getting smaller with every passing year.

Unless we can identify numerous examples of creation that were obviously the work of naturalism, then there are no reasons apart from spiritual reasons, why people believe that the creation gap is like other operation-and-maintenance gaps. The operation-and-maintenance gaps are getting smaller but God is the creator not the operator-and-maintainer. If we do not notice that the creation instance is different to the operation-and-maintenance, then we have no choice but to regard naturalism as probable.

Biochemical machines have extreme complexity that is associated with the purpose of the machine. We can allow that mountain ranges and rivers are created naturally because natural forces can create certain kinds of things. Even for things like the solar system, that do not seem to be examples of extreme complexity, there is no reason to assume that creation of the solar system was the work of operation-and-maintenance laws. This is because the solar system seems to be an example of creation rather than operation-and-maintenance laws.

If we have a faith in philosophical naturalism then the big bang theory seems like a reasonable bet because it had to happen somehow. But, if we do not have a faith in philosophical naturalism, then we need to ask if we would predict anything like our universe on the basis of the theory. We have an expanding universe which is consistent with the big bang theory, but it is also consistent with the idea of a big

God, gravity and the auxiliary assumption that God wants to avoid the big crunch. That means that from the observation of an expanding universe, we cannot extrapolate back to the big bang without the help of the spiritual assumption that the universe was not created.

Although the universe is more visually impressive than the liquid complexity of a biological cell, it may be difficult to argue against the natural creation of the universe because the universe is a symmetrical structure not a machine. A machine, such as a watch, can be described in terms of the laws of the universe and the purpose of the machine, which is a way of saying that part A of the machine predicts part B of the machine, which in turn is a way of saying that we have something that needs explaining. Though symmetry is similar (the left hand predicts the right hand and one part of a circle predicts the rest) symmetry can be part of natural structures. Certain symmetrical surfaces (such as the shapes of the surfaces on Paley's watch, elliptical orbits, and spherical planets) may or may not be difficult to explain within the bounds of naturalism, but to evaluate such ideas we would need to determine what can and what cannot be made by natural processes. The machine argument is not burdened with this problem, because we know that materials do not have the natural propensity to form themselves into machines. However, scientists still need to be able to avoid the religious understanding that 'there must be a natural solution because the planets and stars are up there' and if there is to be any chance of avoiding that understanding they must clearly understand when they are doing their thinking with philosophical naturalism.

On page 4 of their *Science and Creationism*, booklet the National Academy of Sciences mentions that "the background radiation has exactly the correct *spectrum* predicted by the Big Bang origin of the universe". The spectrum correlation suggests there is a lot of independent information that agrees with the theory. However, I think that *NAS* should have said that the spectrum is in agreement with a certain temperature, which is a few degrees above absolute zero, and that this temperature is the temperature (to within some specified accuracy) predicted by the big bang theory. And I think they should have said that the Big Bang theory itself did not predict the uniform background radiation, but it was made consistent with the uniform background radiation by the 'inflation' auxiliary assumption.

Scientific thinking demands that science explain everything in terms of naturalism, because naturalism provides the only scientific explanation. Scientific thinking insists that it must be the case that any natural explanation that half works is a good explanation when there are no alternative natural explanations. Scientists easily dismiss the fact that 'the existence of life is evidence against naturalism', because scientists are thinking with philosophical naturalism.

Johnson is correct — modern science is philosophical naturalism. Contrary to what our spokesmen for evolution argue, there is no scientific point of view outside of philosophical naturalism. Even scientists such as Francis Crick and Fred Hoyle who have agreed that the problem of creation is as creationists see it, have rejected the creationist solution and conjectured that aliens were somehow involved in the creation of life. The Crick-Hoyle alien theory is a strange union of intelligent design and philosophical naturalism.

An intervention miracle is not necessarily highly improbable if we think that we can identify the special circumstances. And having opened the door to non-naturalism, what follows becomes a matter of individual opinion. I do not find the miracles that follow the born again Christian, the witch-doctor, the voodoo priest, the Buddhist priest and perhaps even the martial arts expert and the hypnotist that unbelievable because these are special circumstances. Such circumstances are different to the circumstances of a normal laboratory experiment. On page 433 of *Origin of Species,* Darwin explains his world view:

> If then we have under nature variability and a powerful agent always ready to act and select, why should we doubt that the variations in any way useful to beings, under their excessively complex relations of life, would be preserved, accumulated, and inherited? Why, if man can by patience select variations most useful to himself, should nature fail in selecting variations useful, under changing conditions of life, to her living products? What limit can be put to this power, acting during long ages and rigidly scrutinising the whole constitution, structure, and habits of each creature, — favouring the good and rejecting the bad? I can see no limit to this power, in slowly and beautifully adapting each form to the most complex relations in life. The theory of natural selection, even if we looked no further than this, seems to me to be in itself probable.

Darwin had a different world view, but not all world views are equal. In order for me to object to the beliefs of the Buddhist, I will at some stage need to call on my Scriptural world view and the Buddhist will call on his world view to object to mine. However, I do not need to go anywhere near my Scriptural world view to object to Darwin's argument.

Darwin saw the creation of complexity as probable, given a demonstrable amount of power in artificial selection and his assertion that there were no limits on how far things could evolve under artificial selection. However, Darwin got his confidence for his assertion from philosophical naturalism. Of course, Darwin could see that a natural explanation to life was difficult, but against this he also thought that a natural explanation was probable. In other words, in his mind philosophical naturalism had levelled the playing field and therefore, he only needed relatively weak evidence to be sure that biological machines could be assembled by naturalism. Loosely speaking, we could say that philosophical naturalism understands that the improbability of the natural creation of a sophisticated biochemical machine is balanced out by the improbability of supernatural intervention.

Darwin pointed out the similarities between a human hand, a bats wing, and a whales fin. Steven Gould asked the question 'Why should all the large native mammals of Australia be marsupials, unless they descended from a common ancestor isolated on this island continent?' These are things that a creationists should not expect to find, but we should ask ourselves how probable and how much looking did we do before we found them? And after we sort that out, we should ask ourselves if these are the sort of numbers that we want to use to argue for the natural creation of a sophisticated machine? People who can't see any problems with these kinds of arguments are people who believe that the playing field has been levelled by philosophical naturalism (The notes include a Table of Probability Values that explains the background probability values invoked by philosophical naturalism).

Philosophical naturalism filled the western world during the God-is-dead enlightenment. We are continually being indoctrinated to believe in philosophical naturalism — it is in our schools and universities, in newspapers and magazines, on the television and on the radio, scientists preach it, teachers preach it, journalists preach it, religion preaches it — we are told over and over again by every means possible that there is no evidence for a real God who can do things because he has never done anything. Governments even legislate this kind of thinking in our schools.

Ordinary non-scientific truth says that we are all surrounded by miracles in the form of people, birds, grass and trees and without some spiritual assistance, it seems difficult to understand that there was ever a scientist who failed to notice that creation was not an example of operation-and-maintenance.

Although it is reasonable that scientists might not identify seasons of the earth as a secret cause and time as merely a sensible quality, it is not reasonable that scientists would incessantly and persistently fail to notice that creation is fundamentally different to the problems of operation-and-maintenance of the universe. The unreasonable part of this problem is the spiritual part of this problem.

We can also allow that there is a cultural aspect and an intellectual aspect to the failure to notice that creation is different to maintenance. It could be argued that the western world became over-enlightened to the errors of Aristotles crystalline spheres when we noticed that things could be explained by natural laws that did not include a teleological element.

6.2 Atheism?

Because the ancient Greeks lived in a mysterious world, religious Greeks made their gods the cause of too many things. A few Greeks such as Anaxagoras (~500~428 BC) noticed that many things could be explained by natural causes. According to the *Internet Encyclopedia of Philosophy*: "Anaxagoras of Clazomenae was an important Presocratic natural philosopher and scientist who lived and taught in Athens for approximately thirty years. He gained notoriety for his materialistic views, particularly his contention that the sun was a fiery rock. This led to charges of impiety, and he was sentenced to death by the Athenian court. He avoided this penalty by leaving Athens, and he spent his remaining years in exile." Anaxagoras was more scientific than his fellow Greeks. Regarding living things, Anaxagoras claimed that a substance called 'nous' (mind) entered into dead matter to give it life. Perhaps that means that Anaxagoras believed in a religious form of philosophical naturalism.

As discussed in a note on Russell's Criticism Of Greek Science, Russell argues that Empedocles' (~490~430 BC) philosophy was 'more scientific' than that of Parmenides, Plato, and Aristotle, because it was based on chance and necessity rather than purpose. Russell also argues that the theory of the atomists such as Democritus (and Leucippus) (~460~370 BC) was more nearly like that of modern science than any other theory propounded in antiquity. It would have been better if Russell used a word like 'scientific' to explain Anaxagoras theory that the sun could be explained as a hot rock, and a word like 'enlightened' or 'intellectual' to explain Empedocles' theory of evolution. That is, if Greek science aimed to more sensible rather than scientific, then Greek science would have done better to allow some divine intervention. The idea being that we should use natural law to explain those things that can be explained by natural law, and we should use divine intervention to explain things that cannot be explained by natural law.

The fact that the ancient Greeks used their gods to explain too much may have helped scientific Greeks such as Empedocles to use their naturalism to explain too much. And the fact that Greek religion used their gods to explain too much meant that Greek scientific thinking could be used to undermine Greek religion. On pages 251 to 254 Russell suggests that there were good reasons for Epicurus (~341~270 BC) and his poet disciple, Lucretius (~99~55 BC), to dislike Greek religion. On page 255 Russell concludes:

> The hatred of religion expressed by Epicurus and Lucretius is not altogether easy to understand, if one accepts the conventional accounts of the cheerfulness of Greek religion and ritual. Keats's *Ode on a Grecian Urn*, for instance, celebrates a religious ceremony but not one which could fill men's minds with dark and gloomy terrors. I think popular beliefs were very largely not of this cheerful kind. The worship of the Olympians had less of superstitious cruelty than the other forms of Greek religion, but even the Olympian gods had demanded occasional human sacrifice until the seventh or sixth century B.C., and this practice was recorded in myth and drama. Throughout the barbarian world, human sacrifice was still recognized in the time of Epicurus; until the Roman conquest, it was practised in times of crisis, such as the Punic Wars, by even the most civilized of barbarian populations.
>
> As was shown most convincingly by Jane Harrison, the Greeks had, in addition to the official cults of Zeus and his family, other more primitive beliefs associated with more or less barbarous rites. These were to some extent incorporated in Orphism, which became the prevalent

> belief among men of religious temperament. It is sometimes supposed that Hell was a Christian invention, but this is a mistake. What Christianity did in this respect was only to systematize earlier popular beliefs. From the beginning of Plato's Republic it is clear that the fear of punishment after death was common in fifth-century Athens, and it is not likely that it grew less in the interval between Socrates and Epicurus. (I am thinking not of the educated minority, but of the general population.) Certainly, also, It was common to attribute plagues, earthquakes, defeats in war, and such calamities, to divine displeasure or to failure to respect the omens. I think that Greek literature and art are probably very misleading as regards popular beliefs.
>
> [When Apostle Paul preached to the Epicurean and Stoic philosophers in Berea (see Acts 17:18), the Greeks may have had some understanding of punishment after death. However Hell/sheol is a doctrine of the Mosaic Covenant and the Jews were in contact with surrounding cultures. In the days of Solomon, kings from distant lands came to visit Israel. Furthermore according to Romans 1:21-25 even before the Mosaic covenant, all gentile cultures had some connection to God and therefore all cultures would have had some understanding of hell.]

Epicurus may have had good reason to dislike religion. Perhaps we can add some bitterness and contention to this problem, given that Greek religion attempted to *mandate* that the gods maintained and operated everything.

It is sometimes said that the September 11 terrorism contributed to a rise in atheism. According to a Huffington Post internet article *New Atheist Emerge From 9/11*, authored by Kimberly Winston:

> In September 2001, Sam Harris was an unknown doctoral student who didn't believe in God. But after the World

> Trade Center crumbled on 9/11, he put his studies aside to write a book that became an instant best-seller -- and changed the way atheists, and perhaps Muslims, are perceived in this country.
>
> Published in 2004, Harris's "The End of Faith" launched the so-called "New Atheist" movement, a make-no-apologies ideology that maintains that religion is not just flawed, but evil, and must be rejected...
>
> "I think religion should be treated with ridicule, hatred and contempt, and I claim that right," Hitchens told a Toronto audience in 2007. Freethinkers who are in dialogue with people of faith are "accommodationists," the New Atheists have charged, and "enemies" of the movement. That rift has had real consequences. In 2010, Paul Kurtz resigned as founding leader of the Council for Secular Humanism and the Center for Inquiry in what he described as a "palace coup." Talk amongst the freethinkers was that Kurtz was too accommodationist. "They're anti-religious, and they're mean-spirited, unfortunately," Kurtz told NPR in 2009. "Now, they're very good atheists and very dedicated people who do not believe in God. But you have this aggressive and militant phase of atheism, and that does more damage than good."

Most freethinkers have already undermined their freethinking with philosophical naturalism but these kind of freethinkers have mistaken their Epicurean spleen for their frontal lobe.

Instead of speaking in terms of Islamic terrorism, these kind of freethinkers prefer to speak in terms or religious terrorism. Harris attempts to use the word 'faith' to make his generalisation appropriate but I do not think his argument works very well.

Even if Harris's generalisation is appropriate, it is surely only probable. And when freethinkers say things like 'I think religion should be treated with ridicule, hatred and contempt' and 'that religion is not just flawed, but evil, and must be rejected', then they must surely think that they have discovered some facts that prove that Jesus did not rise from the dead. If people would think with their grey matter instead of their spleens they would see that Islamic terrorism is not even good evidence that Muhammad was not a prophet, because it is possible that God would want unbelievers killed.

If all people who would believe in Jesus could do so despite the influence of the theory of evolution, then may science win the argument and may atheist religion continue to fill our schools and universities and media. May scientists and teachers and religious leaders continue to preach over and over again by every means possible that there is no evidence for a real God who can do things because he has never done anything. If all people who would believe in a real God could do so despite the influence of the theory of evolution, then may Christians continue to be the laughing stock of the intellectual world. However in the real world only about 2% of Australians believe that Jesus rose from the dead, therefore something must be said.

The fact that the Bible says things like 'it is appointed unto man once to die and then the judgement' means that the Bible does not endear itself to freethinkers. Such topics along with the reason people identify terrorism as the evil of religion rather than the evil of Islam, and Bible doctrines are the topics of other books in the *Science Delusion* series. However, this first book does have an interest in how ancient cultures have changed and even if we do not believe the Bible, the Bible still tells us something about ancient cultures.

The New International Version of Job 31:26-28 has Job declaring: "If I have regarded the sun in its radiance or the moon moving in splendor, so that my heart was secretly enticed and my hand offered them a kiss of homage, then these also would be sins to be judged, for I would have been unfaithful to God on high." It doesn't really matter what we think of the book of Job, the book of Job tells us that there was something different about ancient cultures. According to Exodus 19:16-18,20:18-20,32:7-8:

> And it came to pass on the third day in the morning, that there were thunders and lightnings, and a thick cloud upon the mount, and the voice of the trumpet exceeding loud; so that all the people that [was] in the camp trembled.
>
> And Moses brought forth the people out of the camp to meet with God; and they stood at the nether part of the mount. And mount Sinai was altogether on a smoke, because the LORD descended upon it in fire: and the smoke thereof ascended as the smoke of a furnace, and the whole mount quaked greatly. ...
>
> And all the people saw the thunderings, and the lightnings, and the noise of the trumpet, and the mountain smoking: and when the people saw [it], they removed, and stood afar off. And they said unto Moses, Speak thou with us, and we will hear: but let not God speak with us, lest we die.
>
> And Moses said unto the people, Fear not: for God is come to prove you, and that his fear may be before your faces, that ye sin not. ...
>
> And [later] the LORD said unto Moses, Go, get thee down; for thy people, which thou broughtest out of the land of Egypt, have corrupted [themselves]:
>
> They have turned aside quickly out of the way which I commanded them: they have made them a molten calf, and have worshipped it, and have sacrificed thereunto, and said, These [be] thy gods, O Israel, which have brought thee up out of the land of Egypt.

It seems strange that after God delivered Israel from Egypt by bringing 10 supernatural judgements on Egypt, and after

God parted the Red sea and went before the Jews as a pillar of fire by night and a pillar of smoke by day, the Jews decide that they no longer want anything to do with the God of the universe, because Moses was up on the Mountain too long.

The ancients had no understanding of philosophical naturalism. The ancients battled different kind of demons. For the ancients, both Jews and gentiles the problem they wrestled with wasn't 'God versus naturalism' but it was a case of 'god versus god'. Was God just another god? According to the Bible, when Moses cast down his staff and it turned into a crocodile (at the burning bush it became a snake), the Egyptian magicians did the same thing. Moses' crocodile ate up the crocodiles created by the Egyptian magicians. When God took Elijah to heaven, God's prophets went to look through the mountains for Elijah because they thought that God might have dropped Elijah – see 2 Kings 2:16-18. And according to Exodus 18:10-11, when Jethro (Moses' father in law) learned how God delivered the Jews from Egypt: "And Jethro said, Blessed [be] the LORD, who hath delivered you out of the hand of the Egyptians, and out of the hand of Pharaoh, who hath delivered the people from under the hand of the Egyptians. Now I know that the LORD [is] greater than all gods: for in the thing wherein they dealt proudly [he was] above them." Jethro was in relationship with God but apparently he had some lingering doubts about whether his God was just another god.

According to the Bible, ancient Egyptian culture was not the norm either because Jesus said (see Matthew 11:23) that 'if someone had done his mighty works in Sodom that the Sodomites would have repented'. Nevertheless, it seems easy to understand that because the ancients were surrounded by mystery, it was much easier for the ancients to be spiritual. Few ancients were more spiritual than the Canaanites who sacrificed their babies to the god of Molech by burning them alive.

According to the Bible, the judgements on Egypt were God's judgements on gentile religion and the invasion of the land of the Canaanites by the Jews was a judgment on gentile religion. What God did with the ancient Greek culture does not make it into the Bible, but according to the Bible, God distanced himself more and more from the Jews as time went by. According to Exodus 33:2-3, this backing off commenced from the time they invaded the land of the Canaanites: "And I will send an angel before thee; and I will drive out the Canaanite, the Amorite, and the Hittite, and the Perizzite, the Hivite, and the Jebusite: Unto a land flowing with milk and honey: for I will not go up in the midst of thee; for thou [art] a stiffnecked people: lest I consume thee in the way." God backed off a little from the moment they entered the promised land because the Jews had a habit of resisting God's correction. Moses had previously interceded on behalf of the Jews to prevent their judgement, but Moses was not going to be around to intercede for the Jews as they entered the promised land.

During the time of exodus of the Jews from Egypt, God not only did things like turning Moses staff into a crocodile, but God also led the Jews by a pillar of smoke by day and a pillar of fire by night. The fact that God instructed the Jews to tell their children of the supernatural deliverances that God brought to Israel during the time of Exodus means that Exodus was a special time. However it is also true that the Jews turned further and further away from God as time went by. Initially the Jews were ruled by Judges rather than Kings. The book of Judges ends with the statement: "In those days *there was* no king in Israel: every man did *that which was* right in his own eyes." After the Judges came the Kings and the split into the northern and southern tribes. Almost immediately the northern tribes adopted the religions of the Canaanites which resulted in northern Jews being carted off to other lands and being replaced by a people the New Testament refers to as

'Samaritans'. The Jews from the southern tribes followed the northern tribes and adopted the religions of the Canaanites and God used King Nebuchadnezzar to bring judgement on the southern tribes.

Before the Jews were carted off to Babylon by Nebuchadnezzar, Isaiah saw fit to declare (from Isaiah 45:15): "Verily thou [art] a God that hidest thyself, O God of Israel, the Saviour." God brought a great deliverance to the Jews living in Babylon by enabling the Jews as a minority culture living in a foreign land to identify and kill their enemies. Unlike many previous events, such as God parting the Red sea, and God killing every first born Egyptian, the deliverance that God brought to the Jews when they were living in Babylon could be explained in natural terms by Jews predisposed to scientific thinking.

Although it is true that during the time of Exodus, God sometimes used natural means to bring about his plans, the book of Malachi indicates that not long after returning to their land from their captivity in Babylon, the Jews were resisting God's correction. After Babylon, the Jews only had the councils and synagogues instead of prophets because God was hiding himself. From that point on, Judaism becomes more and more religious, except that God remained committed to fulfilling his side of the Mosaic covenant until the time of Jesus when he judged it to be broken.

The New Covenant bears witness to the fact that the Jews did turn almost completely away from God, but unlike previous occasions, they did not turn to idolatry. Unlike previous rebellions against God, neither Jesus nor John the Baptists found it necessary to warn the Jews about idolatry. At the time of Jesus, it would seem that many Jews no longer saw much sense in the warning of Deuteronomy 4:19:

> And lest thou lift up thine eyes unto heaven, and when thou seest the sun, and the moon, and the stars, [even] all

> the host of heaven, shouldest be driven to worship them, and serve them, which the LORD thy God hath divided unto all nations under the whole heaven.

Philosophical naturalism is the spirit of the modern world. Philosophical naturalism protects modern man from idolatry. Acts 23:8 of the Bible says of the Sadducees, "For the Sadducees say that there is no resurrection, neither angel, nor spirit: but the Pharisees confess both." New Testament examples of the contention between the Pharisees and the Sadducees suggests that the Sadducees were comparable in number to the Pharisees. At the time of Jesus, instead of being driven to worship the hosts of heaven, it seems that many Jews were being driven to naturalise their religion. This new kind unbelief could be explained by the fact that the Jews no longer had prophets that could call down fire from heaven because God was hiding himself. It may be that Democritus' and Empedocles' science infiltrated Israel following the conquests of Alexander the Great.

Instead of defining scientific thinking as inappropriate naturalism we could define scientific thinking as thinking based on the understanding that 'there is no God who can do things because if there is a God who could do things then he would be doing things right now'. Or we could say that science is happy to put its faith in superduper naturalism because, 'if there is a God who can do things, then he would not be a God who demands faith'.

The Greeks were amongst the first to use natural law to explain those things that cannot be explained by natural law. A generalisation error explains ancient Greek atomism, ancient Greek evolution and ancient Greek religion. Religious Greeks thought that the gods were responsible for everything. Scientific Greeks thought that natural law was responsible for everything.

It was reasonable that the ancient Greeks thought that they needed gods to explain creation and given the need for such gods, it was not unreasonable to imagine that the gods did in fact operate-and-maintain everything. That is, if we allow that the Greek gods could explain walking, talking, self-reproducing biochemical machines then what could they not do? Although we can say that it was somewhat presumptuous for the religious Greeks to insist that because the gods can create walking, talking, self-maintaining and self reproducing machines, we need the intervention of the gods to explain natural events, it was not unreasonable to suppose that such gods may have been directly involved in natural events.

On the other hand it was both foolish and unreasonable for the scientific Greeks to believe that because the sun could be explained as a hot rock, that creation could be explained in terms of self-organising atoms. That is, although both the religious Greeks and the scientific Greeks failed to distinguish the natural behaviour of the universe from the creation of the universe, we can say that ancient Greek religion was reasonable science and ancient Greek atomism was unreasonable religion. Unreasonable religion believes that there is something more scientific about the kind of thinking that fails to see the difference between the maintenance-and-operation laws and creation.

In the ancient world, the idea that 'there were no gods' was unthinkable. In our enlightened world the idea that 'there is a real God that can do things' is unthinkable because we have natural laws to explain things. Although it seems fair to say that the mystery of the ancient world was undermined by the scientific revolution, the issue is not that we have natural laws to explain those things that can be explained by natural law but that we have super-duper naturalism to explain those things that cannot be explained by natural laws.

Gentile nations may have used gods to explain too much but as discussed in a note on Ancient Attitudes to Naturalism and Intervention it seems that ancient Jews had a reasonable understanding of the difference between naturalism and intervention. The Jews may have regarded the rising and setting of the sun due to the actions of God, but perhaps because the Jews had one God, they did not regard the rising and setting of the sun to the intervention of God. However, Deuteronomy 4:19 tells us that even the Jews were not that far away from identifying those circumstances when their actions might make the sun god angry.

During the transition stage between the Mosaic Covenant and the New Covenant, both Israel and Greece were provinces of the Roman empire. By the end of the first century the Roman empire had spread to all the lands around the Mediterranean Ocean including into Europe as far as the Danube and Rhine rivers and to England and Whales.

The 1 Timothy 6:20-21 Scripture from the New Covenant which followed the Roman empire warned Christians: "O Timothy, keep that which is committed to thy trust, avoiding profane and vain babblings, and oppositions of science falsely so called [gnosis pseudonymos = pseudo-knowledge]: Which some professing have erred concerning the faith. Grace be with thee. Amen." It is possible that this reference to pseudo knowledge was a reference to Greek pseudo-knowledge inspired by philosophical naturalism. Ancient philosophers asserted that their could be no rising from the dead because they thought they had some way of knowing such things. People who believe in philosophical naturalism believe that philosophical naturalism is deducible from observation. This kind of confidence is different to that of faith and revelation.

The Roman empire was more interested in politics and civil and military engineering than in theoretical science and philosophy, and it seems that initially neither Democritus nor Aristotle did so well in the western part (the non-Greek speaking part) of the Roman empire. On page 148 of *The Beginnings of Western Science*, Lindberg explains that by the early 6th century that the Latin west had been largely cut off from original Greek science and natural philosophy. On page 160 Lindberg mentions Themistius (317 -390 AD), Simplicius (490-560) and John Philoponus (490-570 AD) as noteworthy philosophers in the Greek part of the Roman Empire: "And John Philoponus (d. ca. 570), a Christian Neoplatonist who taught in Alexandria, wrote commentaries on Aristotle's *Physics, Meteorology, On Generation* and *Corruption, and On the Soul.* In these commentaries he attempted, in conscious opposition to his contemporary Simplicius, to demonstrate the profound errors propagated by Aristotle, including the celestial-terrestrial dichotomy and the notion of an eternal universe. He also offered a systematic and original refutation of Aristotle's theory of motion, denying Aristotle's explanation of projectile motion and the claim that heavy bodies fall through a medium with speeds proportional to their weights. Through the eventual translation of their works into Arabic and Latin, all three philosophers — Themistius, Simplicius, and Philoponus — helped to determine the subsequent course of Aristotelian natural philosophy."

A little Greek science flowed into Persia via Greeks (such as Simplicius), fleeing the Greek Orthodox Church. Arab Muslims conquered Persia not long after Muhammad died in 632, commencing with raids by caliph Abu Bakr and continuing under caliph Umar and caliph Uthman. Persia, under the influence of an Islamic culture and Arabic language, adopted and advanced the mathematics, science and philosophy, and technology of their neighbours. This included Greek medicine, science and philosophy, Indian mathematics and astronomy and Chinese paper and gun powder. The new Islamic scientific culture spread from

Persia to other parts of the Islamic empire including Muslim conquered Spain, ruled by the Umayyad dynasty.

Al-Gazali (1058-1111) a widely respected fundamentalist Muslim saw the need to write against the philosophers. In the conclusion to his book *The Incoherence of the Philosophers* Gazali writes:

> (1) If someone says: "You have explained the doctrines of these [philosophers]; do you then say conclusively that they are infidels and that the killing of those who uphold their beliefs is obligatory?" we say:
>
> (2) Pronouncing them infidels is necessary in three questions. One of them is the question of the world's pre-eternity and their statement that all substances are pre-eternal. The second is their statement that God's knowledge does not encompass the temporal particulars among individual [existents]. The third is their denial of the resurrection of bodies and their assembly at the day of judgment.
>
> (3) These three doctrines do not agree with Islam in any respect. The one who believes them believes that prophets utter falsehoods and that they said whatever they have said by way of [promoting common] utility, to give examples and explanation to the multitudes of created mankind. This is manifest infidelity which none of the Islamic sects have believed.
>
> (4) As regards questions other than these three, such as their treatment of the divine attributes and their belief in divine unity entailed therein, their doctrine is close to that of the Muctazila. Their doctrine of the necessary [connection] of natural causes [and their effects] is the one which the Muctazila have explicitly expressed in [their doctrine of] generation.

> Similarly, all of what we have reported [of their doctrines] has been held by one or another of the Islamic sects, with the exception of these three principles. Whoever perceives the heretical innovators among the Islamic sects to be infidels will pronounce [the philosophers] also to be infidels [for the views they share]. And whoever hesitates in pronouncing [such Islamic sectarian innovators] as infidels will confine himself to pronouncing [the philosophers] infidels on these [three] questions [alone]. We, however, prefer not to plunge into [the questions] of pronouncing those who uphold heretical innovation to be infidels! and of which pronouncement is valid and which is not, lest the discourse should stray from the objective of this book. God, exalted be He, is the One who leads one successfully to what is correct.

Muslim philosophers asserted that there could be no rising from the dead. Philosophers had no way of knowing about such things without the help of a religious doctrine like philosophical naturalism.

Science and religion can overlap, but that does not mean that religion is anti-science. Ancient cultures such as the Chaldeans, Hindus and Sabians had a religious focus on the stars, and this may have been important to the early development of mathematics and astronomy. Genesis 11:4 says: "And they said, Go to, let us build us a city and a tower, whose top [may reach] unto heaven; and let us make us a name, lest we be scattered abroad upon the face of the whole earth." The words in square brackets have been inserted by the translators. According to F. J. Dake's note on page 9 of *Dake's Annotated Reference Bible*, "The meaning of this is that the top would be dedicated to the heavens, with signs of the zodiac portrayed on it as visible objects of worship."

Arabic mathematics, science and technology eventually passed to the western Europe. In the early 13th century, translations of Arabic works including commentaries on Aristotle from Avicenna (Ibn Sina, 980- 1037) and Averroes (Ibn Rushd, 1126-98) entered western Europe mainly from Spain. Although Averroes refers to Aristotle as 'the Philosopher and the Master of them that know' Russell argues for a general decay in Greek science after Democritus. On page 90 of his *History of Western Philosophy* Russell writes: "In spite of the genius of Plato and Aristotle, their thought has vices which proved infinitely harmful. After their time, there was a decay of vigour, and a gradual recrudescence of popular superstition. A partially new outlook arose as a result of the victory of Catholic orthodoxy; but it was not until the Renaissance that philosophy regained the vigour and independence that characterize the predecessors of Socrates" When Russell speaks of the new outlook provided by Catholic orthodoxy, he perhaps means that the Catholic Church opposed some Greek science and philosophy using the understanding that God can do what he likes in the empirical world.

Not only did ancient Greek science say that the world had no beginning but it also said the sun, planets, and stars revolved around the earth and such a geometry made it difficult for the ancients to account for the fact that the planets sometimes appear to stop and move backwards for some time before moving forwards again. Instead of the planets being fixed to celestial spheres, the planets were thought to perform small orbits (epicycles or circles superimposed on circles) around points fixed to a main deferent sphere. Although this retrograde motion was included in Ptolemy's mathematical model, the scientific culture of the time required that astronomers find an accurate mechanical model based on circles.

Many people mark the beginning of the scientific revolution with Copernicus' sun centred theory of the solar system. Copernicus simplified the model of the solar system by putting the sun rather than the earth at the centre. On page 513 of *History of Western Philosophy*, Russell writes of Copernicus (1473-1543):

> He came early to believe that the sun is at the centre of the universe, and that the earth has a twofold motion: a diurnal rotation, and an annual revolution about the sun. Fear of ecclesiastical censure led him to delay publication of his views though he allowed them to become known. His chief work, *De Revolutionibus Orbium Crelestium*, was published in the year of his death (1543), with a preface by his friend Osiander saying that the heliocentric theory was only put forward as a hypothesis. It is uncertain how far Copernicus sanctioned this statement, but the question is not very important, as he himself made similar statements in the body of the book. The book is dedicated to the Pope, and escaped official Catholic condemnation until the time of Galileo. The Church in the lifetime of Copernicus was more liberal than it became after the Council of Trent, the Jesuits, and the revived Inquisition had done their work.
>
> The atmosphere of Copernicus's work is not modem; it might rather be described as Pythagorean. He takes it as axiomatic that all celestial motions must be circular and uniform, and like the Greeks he allows himself to be influenced by aesthetic motives. There are still epicycles in his system, though their centres are at the sun, or, rather, near the sun. The fact that the sun is not exactly in the centre marred the simplicity of his theory. Though he had heard of the Pythagorean doctrines, he does not seem to have known of Aristarchus's heliocentric theory, but there is nothing in his speculations that could not have occurred to a Greek astronomer. What was important in his work was the dethronement of the earth from its geometrical preeminence. In the long run, this made it

> difficult to give to man the cosmic importance assigned to him in the Christian theology, but such consequences of his theory would not have been accepted by Copernicus, whose orthodoxy was sincere, and who protested against the view that his theory contradicted the Bible...

Modern scholars tell us that hundreds of years before Copernicus, Ibn al-Shatir (1304 – 1375) proposed a model of the solar system which employed Nasiral-Din al-Tusi's (1201-- 1274) Tusi couple and that the mathematics of the Shatir's solution was similar to the Copernicus (1473 – 1543) model except that Copernicus flipped the geometry around and allowed the sun to be at the centre.

That the Koran seems to support ancient Greek astronomy, may have made it difficult for Muslim astronomers to do what Copernicus did. In response to Copernicus' science, Luther asserted that Joshua commanded the 'sun to stand still and not the earth', however, Joshua 10:12-13 says: "Then spake Joshua to the LORD in the day when the LORD delivered up the Amorites before the children of Israel, and he said in the sight of Israel, Sun, stand thou still upon Gibeon; and thou, Moon, in the valley of Ajalon. And the sun stood still, and the moon stayed, until the people had avenged themselves upon their enemies. [Is] not this written in the book of Jasher? So the sun stood still in the midst of heaven, and hasted not to go down about a whole day." Luther would have had a real argument if Joshua ignored the moon (because Joshua required more daylight) and as a consequence of such a command, the sun stood still and the moon kept moving.

Russell says that putting the earth at the centre was important to Christian theology. I am not sure if Russell means in a fuzzy religious way of if he means that all mechanical models of the solar system and even the understanding that the earth was the centre of the universe included a teleological element.

Before the scientific revolution, man thought that the heavens were perfect and unchanging compared to the earth which was seen as corrupt and decaying. The understanding of a perfect and unchanging heavens was challenged by observation. Tycho Brahe observed the formation of a new star and following the invention of the telescope, Galileo (1564-1642) observed spots on the sun and craters on the moon. The understanding of an imperfect heavens made it easier for scientists to consider a universal principle of repeatability (because the earth was no longer a special place). And the understanding of a moving earth made it easier for scientists to think in terms relative motion and inertia.

Johannes Kepler (1571 – 1630), perfected Copernicus model with the use of elliptical orbits. Although Kepler retained a celestial sphere to support the stars he offered no mechanical solution to explain the movement of the planets. In the absence of a mechanical solution, Descartes (1596-1650) suggested that a whirlpool of invisible particles impacted on and moved the planets.

Although we should be more impressed with the purpose in the design of life than the purpose in a few crystalline spheres, it seems that the removal of purpose from the design of the heavens added to the confidence of freethinkers. After Galileo, but before Newton published his *Principia*, Benedict Spinoza (1632-1677) wrote in *The Chief Works of Benedict de Spinoza, volume 1 (Tractatus- Theologico-Politicus, Tractatus Politicus)* Chapter VI — 'Of Miracles':

> We may, then, be absolutely certain that every event which is truly described in Scripture necessarily happened, like everything else, according to natural laws; and if anything is there set down which can be proved in set terms to contravene the order of nature, or not to be deducible therefrom, we must believe it to have been foisted into the sacred writings by irreligious hands;

> for whatsoever is contrary to nature is also contrary to reason, and whatsoever is contrary to reason is absurd, and, ipso facto, to be rejected.

Spinoza seems to have in mind mathematically certain knowledge even though he seems to believe in a religious form of philosophical naturalism. Perhaps this was due more to his words-and-definitions thinking than to philosophical naturalism. Spinoza seems to have combined Descartes' rationalism and Hobbes' understanding of familiarity to deduce an extremely confident understanding. Spinoza argued that miracles are those things which we do not yet understand. Patterns that we learn from experience become part of our 'innate' knowledge and like modern day science, Spinoza did not understand that his confidence was inspired by his failure to distinguish creation from maintenance.

Spinoza was ahead of his time because he came before Newton and Darwin. Newton's *Principia* (1687) used and established the principle of repeatability. The new law of gravity that held the planets in their orbit was the same law that made things heavy on the earth. Newton's third law required that if the sun exerted an attractive force on a planet, then the planet exerted an equal force on the sun. The old system said that celestial bodies were made of a non-earthly material and therefore it was difficult to think in terms of universal laws, whereas, Newton's law of gravitation does not distinguish one mass from the other or one location from another. New thinking said that the earth was just a planet like so many other planets and that the sun was just a star like so many other stars. On page 422 of *Never at Rest: A Biography of Isaac Newton*, Westfall writes:

> The satellites of Jupiter were so many more celestial pendulums. Not only did their conformance to Kepler's third law reveal the proportionality of their masses to their attractions toward Jupiter, but their concentric orbits

> about Jupiter demonstrated that the sun attracts them both and Jupiter in proportion to mass. According to Newton's third law the satellites that must attract Jupiter must attract the sun in return. Apparently every body in the world attracts every other body. We can only imagine the wild surmise that raced through Newton's mind as the principle of universal gravitation silently unfolded.

The simplicity and the predictive power of Newton's laws provided confidence in Newton's science. The idea that the planets moved about the sun under natural law rather than due to the action of celestial spheres provided a new way of looking at the universe. Mathematical laws replaced closed mechanical explanations.

Prior to Newton, it must have seemed as if an ad hoc explanation was required to explain the fact that bodies like the moon did not fall to the earth given that rocks did fall to the earth. Using Aristotle's thinking we might expect to see a teleological element specifically for the purpose of supporting the planets given that the eye, the ear and the arm seem so obviously designed for doing the things that they do.

Mechanical explanations seem to include a teleological element and an ad hoc element. Kepler made mechanical explanations difficult. Newton made ad hoc explanations redundant.

A creationist would say that the creator turned out to be smarter than we thought because he does not need additional laws or mechanisms to keep the planets in their orbits. However, freethinkers became convinced that because there is no evidence of a creator in the laws that move the planets, then there is no evidence of a creator in biochemical life. Baron Paul-Henri Dietrich Holbach (1723 - 1789), a contemporary of Laplace, and an acquaintance of David Hume shows us how this sort of freethinking works. Unlike Spinoza, Holbach did not use words-and-definitions assertions and that means that his arguments still make sense today.

Holbach published his *System of Nature* in French in 1770 under the name of Jean-Baptiste de Mirabaud. The English version used in the following notes is that translated by Samuel Wilkinson which is available on the *Free Thought Archives.* Another version translated by H. D. Robinson is available at Google Books and McMaster University.

Unlike Spinoza, Holbach seemed to be a God hater in the spirit-and-motive sense. There are things that Holbach wrote about God that should not be read. On the other hand, unlike so many religion friendly scientists, he was open and honest. Holbach does not seem concerned about letting his audience understand that his confidence comes from philosophical naturalism. In part 2 chapter XI of his *System of Nature*, Holbach, writes: "It is true, perhaps, that the system of naturalism hitherto has not been developed in all its extent: unprejudiced persons however, will, at least, be enabled to know whether the author has reasoned well or ill; whether or not he has attempted to disguise the most important difficulties; distinctly to see if he has been disingenuous; they will be competent to observe if, like unto the enemies of human reason, he has recourse to subterfuges, to sophisms, to subtle discriminations, which ought always to make it suspected of those who use them, either that they do not understand or else that they fear the truth." As illustrated in the following citations Holbach gets his confidence from the observation that all of God's operation-and maintenance laws are natural. Holbach was convinced that philosophical naturalism was manifest truth and that God thinking was delusional:

- [part 1 chapter I] MAN has always deceived himself when he abandoned experience to follow imaginary systems.--He is the work of nature.--He exists in Nature.--He is submitted to the laws of Nature.--

He cannot deliver himself from them:--cannot step beyond them even in thought. It is in vain his mind would spring forward beyond the visible world: direful and imperious necessity ever compels his return--being formed by Nature, he is circumscribed by her laws; there exists nothing beyond the great whole of which he forms a part, of which he experiences the influence. The beings his fancy pictures as above nature, or distinguished from her, are always chimeras formed after that which he has already seen, but of which it is utterly impossible he should ever form any finished idea, either as to the place they occupy, or their manner of acting--for him there is not, there can be nothing out of that Nature which includes all beings. [Instead of saying that naturalism is everything Holbach could have said that 'creationism is not science'.]

- [part 1 chapter II] If Nature had been viewed uninfluenced by prejudice, they must have been long since convinced that matter acts by its own peculiar activity; that it needs no exterior communicative force to set it in motion. They might have perceived that whenever mixed bodies were placed in a situation to act on each other, motion was instantly excited; and that these mixtures acted with a force capable of producing the most surprising results.

If particles of iron, sulphur, and water be mixed together, these bodies thus capacitated to act on each other, are heated by degrees, and ultimately produce a violent combustion. If flour be wetted with water, and the mixture closed up, it will be found, after some lapse of time, (by the aid of a microscope) to have produced organized beings that enjoy life, of which the water and the flour were believed incapable: it is thus that inanimate matter can pass into life, or animate matter, which is in itself only an assemblage of motion.

Reasoning from analogy, which the philosophers of the present day do not hold incompatible, the production of a man, independent of the ordinary means, would not be more astonishing than that of an insect with flour and water. Fermentation and putrid substances, evidently produce living animals. We have here the principle; with proper materials, principles can always be brought into action. That generation which is styled uncertain is only so for those who do not reflect, or who do not permit themselves, attentively, to observe the operations of Nature.

The generative of motion, and its development, as well as the energy of matter, may be seen everywhere; more particularly in those unitions in which fire, air, and water, find themselves combined. These elements, or rather these mixed bodies, are the most volatile, the most fugitive of beings; nevertheless in the hands of Nature, they are the essential agents employed to produce the most striking phenomena. To these we must ascribe the effects of thunder, the eruption of volcanoes, earthquakes, &c. Science offers to our consideration an agent of astonishing force, in gunpowder, the instant it comes in contact with fire. In short, the most terrible effects result from the combination of matter, which is generally believed to be dead and inert.

These facts prove, beyond a doubt, that motion is produced, is augmented, is accelerated in matter, without the help of any exterior agent: therefore it is reasonable to conclude that motion is the necessary consequence of immutable laws, resulting from the essence, from the properties existing in the different elements, and the various combinations of these elements. Are we not justified, then, in concluding, from these precedents, that there may be an infinity of other combinations, with which we are unacquainted, competent to produce a great variety of motion in matter, without being under the necessity of having recourse, for the explanation, to agents who are more difficult to comprehend than even the effects which are attributed to them? [Holbach has a problem with God thinking but no problem with magical thinking]

- [part 1 chapter IV] When any extraordinary, any unusual, effect is produced, to which our eyes have not been accustomed; or when we are ignorant of the energies of the cause, the action of which so forcibly strikes our senses, we are tempted to meditate upon it, and take it into our consideration. The European, accustomed to the use of GUNPOWDER, passes it by, without thinking much of its extraordinary energies; the workman, who labours to manufacture it, finds nothing marvellous in its properties, because he daily handles the matter that forms its composition. The American, to whom this powder was a stranger, who had never beheld its operation, looked upon it as a divine power, and its energies as supernatural. The uninformed, who are ignorant of the true cause of THUNDER, contemplate it as the instrument of divine vengeance. The experimental philosopher considers it as the effect of the electric matter, which, nevertheless, is itself a cause which he is very far from perfectly understanding.--It required the keen, the penetrating mind of a FRANKLIN, to throw light on the nature of this subtle fluid--to develop the means by which its effects might be rendered harmless-- to turn to useful purposes, a phenomenon that made the ignorant tremble--that filled their minds with terror, their hearts with dismay, as indicating the anger of the gods:

impressed with this idea, they prostrated themselves, they sacrificed to JUPITER, to deprecate his wrath.

Be this as it may, whenever we see a cause act, we look upon its effect as natural: when this cause becomes familiar to the sight, when we are accustomed to it, we think we understand it, and its effects surprise us no longer. Whenever any unusual effect is perceived, without our discovering the cause, the mind sets to work, becomes uneasy; this uneasiness increases in proportion to its extent: as soon as it is believed to threaten our preservation, we become completely agitated; we seek after the cause with an earnestness proportioned to our alarm; our perplexity augments in a ratio equivalent to the persuasion we are under: how essentially requisite it is, we should become acquainted with the cause that has affected us in so lively a manner. As it frequently happens that our senses can teach us nothing respecting this cause which so deeply interests us--which we seek with so much ardour, we have recourse to our imagination; this, disturbed with alarm, enervated by fear, becomes a suspicious, a fallacious guide: we create chimeras, fictitious causes, to whom we give the credit, to whom we ascribe the honour of those phenomena by which we have been so much alarmed. It is to this disposition of the human mind that must be attributed, as will be seen in the sequel, the religious errors of man, who, despairing of the capacity to trace the natural causes of those perplexing phenomena to which he was the witness, and sometimes the victim, created in his brain (heated with terror) imaginary causes, which have become to him a source of the most extravagant folly

In Nature, however, there can be only natural causes and effects; all motion excited in this Nature, follows constant and necessary laws: the natural operations, to the knowledge of which we are competent, of which we are in a capacity to judge, are of themselves sufficient to enable us to discover those which elude our sight; we can at least judge of them by analogy. If we study Nature with attention, the modes of action which she displays to our senses will teach us not to be disconcerted by those which she refuses to discover. Those causes which are the most remote from their effects, unquestionably act by intermediate causes; by the aid of these, we can frequently trace out the first. If in the chain of these causes we sometimes meet with obstacles that oppose themselves to our research, we ought to endeavour by patience and diligence to overcome them; when it so happens we cannot surmount the difficulties that occur, we still are never justified in concluding the chain to be broken, or that the cause which acts is SUPER-NATURAL. Let us, then, be content with an honest avowal, that Nature contains resources of which we are ignorant; but never let us substitute phantoms, fictions, or imaginary causes, senseless terms, for those causes which escape our research; because, by such means we only confirm ourselves in ignorance, impede our enquiries, and obstinately remain in error. [Hobach uses Hobbes' concept of familiarity to argue that the only kind of miracles are those that we do not yet understand. Holbach argues that all gaps in our knowledge are the same as gaps in knowledge about operation and maintenance issues.]

- [part 1 chapter VI] Thus, his [man's] imagination, no longer guided by experience, hurried on by his new ideas, was lost, without hope of return, in the labyrinth of an ideal, of an intellectual world, to which he had himself given birth; it was next to impossible to withdraw him from this delusion, to place him in the right road, of

> which nothing but experience can furnish him the clue. Nature points out to man, that in himself, as well as in all those objects which act upon him, there is never more than matter endowed with various properties, diversely modified, that acts by reason of these properties: that man is an organized whole, composed of a variety of matter; that like all the other productions of Nature, he follows general and known laws, as well as those laws or modes of action which are peculiar to himself and unknown. [Holbach fails to notice that creation is different to operation and maintenance. This little generalisation error is sufficient to cause Holbach to assume that there was a natural solution to the creation of life by spontaneous generation.]

The ancient Greeks and the Sadducees had already seen the same light, but Holbach's considered his version to be so much brighter to people who could see the operation-and-maintenance mystery falling away. See Professor Norman Geisler's article titled *Miracles and Modern Scientific Thought* for more examples of scientific thinking.

Under Note 129 at the bottom of volume 1 of the Robinson translation of Holbach's *System Of Nature*, Robinson suggests 'that the larger animals we now behold were originally derived from the smallest microscopic ones' apparently because by that time everyone was catching on to the idea of using natural law to explain those things which cannot be explained by natural law. Robinson's translation was published the year before Darwin's *Origin of Species* was published. From the Preface of the Editor's (J. W. Burrow's) Introduction to Darwin's *Origin of Species*. From p 218 of Adrian Desmond's and James Moore's *Darwin*, we can see Holbach's thinking in Darwin's science and modern science:

> Darwin had asked in his 1842 sketch, comparing the state of biology with physics, "What would the Astronomer say

> [chuckle, chuckle] to the doctrine that the planets moved [not] according to the laws of gravitation, but from the Creator having willed each separate planet to move in its particular orbit". After 1859 biologists no longer needed to say things of that kind and nor did anyone else.
>
> These discussions took place as Darwin was pondering the divine government himself. He too came to accept that 'the Creator creates by...laws.' Law ruled the earth, as it did the heavens; anything else was demeaning of God. Darwin complained that 'we can allow the satellites, planets, suns, universe, nay whole systems of universe[s,] to be governed by laws, but the smallest insect, we want to be created at once by a special act.' It was absurd; just as the winds in the Andes obeyed regular laws, so did the comings and goings of animals on the planet's surface. Dining at Leyell's, dancing at Babbage's, he found the idea of miraculous, catastrophic interruptions increasingly deplored. The rule of law had to be upheld.

Darwin understood his science to be a natural extension of the Newton's principle of using natural law to explain operation-and-maintenance mystery. Burrow, Desmond, and Moore (and Darwin) fail to recognise that under the creationist model, all operation-and-maintenance problems will be solved by naturalism because God has established laws of the universe so that he does not need to personally supervise every particle and every photon in the universe.

Darwin (and his supporters) are comparing the theory of evolution (a theory of creation) with Newton's laws without noticing that Newton's laws are operation-and-maintenance laws. In the above citation Burrow asks what the astronomer would say. In his *Principia*, Book 3, p 440 after raising some more objections to the idea of vortices to explain the orbits of the planets what *The Astronomer* says could hardly be more different to what Burrow, Desmond, Moore and Darwin had in mind:

> Bodies projected in our air suffer no resistance but from the air. Withdraw the air, as is done in Mr. *Boyle's* vacuum, and the resistance ceases; for in this void a bit of fine down and a piece of gold descend with equal velocity. And parity of reason must take place in the celestial spaces above the earth's atmosphere; in which spaces, where there is no air to resist their motions, all bodies will move with the greatest of freedom; and the planets and comets will constantly pursue their revolutions in orbits given in kind and position, according to the laws above explained; but though these bodies may, indeed, preserve in their orbits by the mere laws of gravity, yet thy could by no means have at first derived the regular position of the orbits themselves from those laws.
>
> The six primary planets are revolved about the sun in circles concentric with the sun, and with motions directed towards the same parts, and almost in the same plane. Ten moons are revolved about the earth, Jupiter and Saturn, in circles concentric with them, with the same direction of motion, and nearly in the planes of the orbits of those planets; but it is not to be conceived that mere mechanical causes could give birth to so many regular motions,...

If it was the success of things like Newton's laws that sealed our enlightened belief in naturalism-only, it did not come from Newton, because Newton understood the difference between creation and operation-and-maintenance. People like Holbach and Darwin used the success of Newton's science to replace an omnipotent God with super-duper naturalism because their world view prevented them from distinguishing creation from operation-and-maintenance.

Australian children grow up with the understanding that we do not need a creator to explain creation. Australian children grow up with the understanding that slow changes, no matter how miraculous are natural, because science lends it's support to Holbach's vile hatred for God.

Science zigged when it should have zagged. The notes includes some Creationists arguments repeated in Holbach's System of Nature (first published in 1770), that still work today.

If we suppose that Holbach's misunderstanding is a part of modern science, then even if there was no evidence for the theory of evolution, we should expect modern science to insist that the theory of evolution can explain creation. In other words given the delusion of philosophical naturalism and no evidence for evolution, we should expect the track record of science to look something like what we have got.

Jumbo Jets might be able to fly, mobile phones might work even though they seem incredible. If there is enough time available or if the universe is big enough then, who knows what could happen? Of course, I do not believe that, but short of God coming out of hiding, it might prove impossible to dent Holbach's extreme confidence in philosophical naturalism and western culture is always going to be listening the voice of science over the voice of Christian fundamentalists. Therefore the next best thing is to get Holbach to speak the ordinary unscientific truth about the problem of creation, regardless that he believes that such truth is misleading truth

6.3 Truth

On page 8 of their *Science and Creationism* booklet, The National Academy of Sciences writes:

> The arguments of creationists are not driven by evidence that can be observed in the natural world. Special creation or supernatural intervention is not subjectable to meaningful tests, which require predicting plausible results and then checking these results through observation and experimentation. Indeed, claims of "special creation" reverse the scientific process. The explanation is seen as unalterable, and evidence is sought only to support a particular conclusion by whatever means possible.

It is true that we creationists are not scientific because we are so intent on defending creationism. We creationists do not want to preach the short comings of creationism from the roof tops. However, I cannot see how the understanding that 'we creationists seek to support creationism by whatever means possible' is of much help to creationists or to NAS readers. The issue here is that, such statements can be used like 'religion has been responsible for many wars therefore Jesus did not rise from the dead' and 'creationism is not testable therefore it is not science' as if there are no creationist arguments that should be considered by science.

The *Creation Ministries International* (CMI) site discusses some arguments for creation. Although I have not read all that the CMI site has to say about evolution and creationism, what I have read seems to be good science, good attitude and good thinking. These are arguments that should be considered by science.

CMI also lists many arguments that creationists have used that they should not use. However instead of saying that creationists reverse the scientific process CMI says: "The primary authority for Creation Ministries International is the infallible Word of God, the Bible. All theories of science are fallible, and new data often overturn previously held theories." I like the way CMI put it because it lets us know that 'yes creationism is based on the fixed doctrines of the bible' but words-and-definitions cannot be used to tell us that there is no such thing as a creator. Creationists were once stuck on the idea that the earth had a beginning, and scientists were once stuck on the idea that the earth did not have a beginning. In time science came to agree that the earth had a beginning, which is good, because when we are wrong we need to change.

However it is important to remember that not all theories of a science are alterable. Scientific thinking believes in infinite universes, clay crystals and evolution compatible complexity because the understanding that 'there is no God who can do things' is an unalterable doctrine of science. And because we know from Paley's argument that there never will be an empirical law of the universe that explains creation, we know that science is stuck on a logical solution that must include a trial and error algorithm with a feedback element.

The doctrine of philosophical naturalism is not absolutely unalterable because it will be altered when God comes out of hiding in the not too distant future. A degree of reformation may be possible before that, if science can bring itself to criticise arguments that end up in the scientific place.

Truth is one of the many themes of the book of *Job*. The book of *Job* is the oldest book in the Bible, written by Job when the Jews were slaves in Egypt.

Job hears the news that some of his servants have been killed and oxen and asses stolen by Sabean people, 7000 of his sheep and servants have been killed by fire from heaven, more of his servants killed and 3000 camels stolen by Chaldean people and that his sons and daughters have been killed in a house which was destroyed by a strong wind. Job receives all this bad news over a short space of time. A little later, in a second round of troubles, Job gets a very bad case of boils from the sole of his foot to his head, is afflicted with terrifying nightmares, and his wife leaves him. Job, in his despair, is visited by three friends Eliphaz, Bilad and Zophar and Job finds himself defending his case against God to his three friends, claiming that he is innocent of sin against God.

Job maintained his innocence, and he maintained that if he did something wrong that God should forgive him. According to Job 7:11-19, Job argued that God's ways are not just: "Therefore I will not refrain my mouth; I will speak in the anguish of my spirit; I will complain in the bitterness of my soul. Am I a sea, or a whale, that thou settest a watch over me? When I say, My bed shall comfort me, my couch shall ease my complaint; Then thou scarest me with dreams, and terrifiest me through visions: So that my soul chooseth strangling, and death rather than my life. I loathe it; I would not live always: let me alone; for my days are vanity. What is man, that thou shouldest magnify him? And that thou shouldest set thine heart upon him? And that thou shouldest visit him every morning, and try him every moment? How long wilt thou not depart from me, nor let me alone till I swallow down my spittle? I have sinned; what shall I do unto thee, O thou preserver of men? why hast thou set me as a mark against thee, so that I am a burden to myself? And why dost thou not pardon my transgression, and take away mine iniquity? for now shall I sleep in the dust; and thou shalt seek me in the morning, but I [shall] not [be]."

Job's friends argued that regardless of Job's outward appearance and his good deeds that Job must have had some secret sin in his life because it was obvious that the things happening to Job were supernatural. Eliphaz commenced his arguments with Job 4:7-8: "Remember, I pray thee, who ever perished, being innocent? or where were the righteous cut off? Even as I have seen, they that plough iniquity, and sow wickedness reap the same."

As the counselling session progressed, Eliphaz continued his argument along the same lines, of defending God and condemning Job. Job 22:5-6 has Eliphaz saying to Job: "Is not thy wickedness great? and thy iniquities infinite? For thou have taken a pledge from thy brother for nought, and stripped the naked of their clothing." And Job 11:5-6 has Zophar saying something similar: "But oh that God would speak, and open his lips against thee; And that he would shew thee the secrets of wisdom, that they are double to that which is! Know therefore that God exacteth of thee less than thine iniquities deserveth."

According to Job 19:23 and Job 23:3-4, Job requested: "Oh that my words were now written! oh that they were printed in a book! ... Oh that I knew where I might find him! that I might come even to his seat! I would order my cause before him, and fill my mouth with arguments." At the end of these long arguments between Job and his counsellors, Job got both of his requests. His arguments were written in a Book. And God appeared before Job and challenged him, Job 38: 2-3: "Who is this that darkeneth counsel with words without knowledge? Gird up thy lions like a man: for I will demand of you and you will answer me." God attacked Job's confidence in his wisdom and understanding. Job 42: 7-8 tells us that after God had dealt with Job, that God then dealt with Job's friends: "And it was so that after the Lord had spoken these words unto Job, the Lord said to Eliphaz the Temanite, My wrath is kindled against you, and against your two friends: for you have not spoken of me the thing that is right, as my servant Job hath. Therefore take unto you now seven bullocks and seven rams and go to my servant Job and offer up for yourselves a burnt offering; and my servant Job shall pray for you: lest I deal with you after your folly, in that you have not spoken of me the thing that which is right, like my servant Job."

One of the many purposes of the book of Job is to explain how trials work in the lives of people in relationship with God — the book opens with a discussion that God and Satan are having about Job, with God setting up Job up for a trial by Satan. God wanted the best for Job, but the trial itself required that Satan's role was concealed so that Job's thinking about God could be exposed and corrected. Job thought that God hid thoughts about him in his heart. Job was half right because God knew that there was something wrong in Job's heart, and he secretly determined to work that out so that he could bless Job greatly. Job never considered that he had secret thoughts in his heart about God that weren't so good, because he agreed with his own thoughts. Job doubted God's character- see Job 1:5,22; 7:20-21; 10:3-6, 13, 14:16 and even thought that God helped the wicked: Job 12:6. Job was righteous in his own eyes: Job 32:1. Job trusted too much in his own wisdom and thought himself to be an instructor of God: Job 38. Job did not have a good appreciation for God: Job 38:18. He justified himself and condemned God: Job 40:8. And on the other hand, Job's friends took it upon themselves to find out and judge Job's sins, and they seem to take offence at Job's sins: Job 19: 4, 22 & 28.

Regarding the issue of truth the lesson of Job is that Eliphaz and his friends were out to bat for God (the-Truth), with the result that God was considering some sort of judgement on them. This is why Job asked his friends (Job 13:7): "Will you speak wickedly for God? and talk deceitfully for him?" Although it may be easy to believe that, it is not possible to be on the wrong side if we are arguing from the principle that we will only speak in favour of God and his kingdom (or some other noble truth), the book of Job shows us that this is not the case. God was considering judgement on the people who argued for the-truth of all truths. On the other hand, Job was saying nasty things about the-truth of all truths, and only repents in response to a direct confrontation with God, God considers Job's arguments as more truthful than those of his friends.

There is the-truth that is the God's-eye-view of reality or the objective of the argument, which we aim to investigate, discover or debate. Christians and Scripture itself, regards Scripture as the-truth. Scientists believe that evolution is the-truth. We attempt to investigate and discover the-truth, but we may never be able to say for certain what is and what is not the-truth. However, there is another kind of truth because the New Testament sometimes refers to the Holy Spirit as the spirit of truth. Proverbs 3:3-4 says: "Let not mercy and truth forsake thee: bind them about thy neck; write them on the tablet of your heart..." and because 'truth' in this Scripture is placed alongside 'mercy' (or loving-kindness) then we can interpret this as being the spirit of truth or an attitude of truth. This spirit of truth or attitude of truth (or truthfulness) is different from say, the truth of science or the truth of scripture. This kind of truth is an unbiased arbitrator in an argument. In theory, this truth has an independent voice, which cannot be enlisted to the cause of God, faith, religion, science or justice and people call on this arbitrator to settle issues in these areas.

This understanding means that our part to play in the game of truth has only to do with the methods we use to demonstrate and validate issues of truth that have something to say about the-truth. If we disobey the rules of logic or mathematics or if we use rhetoric to mislead then we are not playing by the rules of truth. This is the type of truth that we have control over and it is the spirit-of-truth. Although the objective in our arguments is always the first kind of truth, our only opportunity for submission to truth is in our arguments and reasoning not in whether we are right or wrong. If we miss this opportunity, regardless of whether we are on the side of the-truth, then our spirit-and-motive is not truth. Whether we are right or wrong is irrelevant to the spirit-of-truth.

The reason why we need to identify these two types of truth is because *we live in an imperfect world.* As a fundamentalist, I may need to say 'I do not find fault with the argument but I do not accept the argument' because truth can be misleading. Truth can be dangerous. Such things make it easy to overstep the mark, since the objective in the end is the-truth (or the-purpose), and the spirit-of-truth is merely a means of getting to our objective. A scientist might deny that there are any problems with the theory of evolution or in another setting, he might be tempted to adjust the results of his experiment, not because he has some sinister plan to deceive the world but because he knows that there are measurement inaccuracies and other problems that can conceal the true results. The theologian might invent some doctrine to help out God's people, because he believes the bible is hard to understand and because an ordinary interpretation of Scripture doesn't seem to be good for Christians or unbelievers. And, an officer of the law, might do the legal system a service (say by elimination of some inconsistent evidence) not because he has any desire to send an innocent person to gaol, but to ensure that the guilty person does not escape justice.

It is because we live in an imperfect world that we can justify pushing the objective (the-truth) further and further away from the spirit-of-truth. For the legal officer, theologian, and scientist the objective is to do the truth a service. This is why Job 13:7 has Job asking his friends: "Will you speak wickedly for God? and talk deceitfully for him?" In our dreams we would never do that, but in real life we would if the truth is important.

To Job's friends, it was obvious that the calamities that struck Job were supernaturally empowered. It may have also seemed obvious to Job's friends that the pattern in Job's life was set — Job had lost everything including his health and it seemed as if he was heading towards the grave. Job said some nasty things about God and if that wasn't sufficient to guarantee the understanding that Job deserved what he was getting, he also said some nasty things about his counsellors. If we can get out of our dreams, we should be able to see that the argument against Job would have been compelling when mixed with a little irresponsibility. Job was accusing God and defending himself. The calamities were obviously supernatural. In other words Job's counsellors got their confidence from a prejudgement and they thought they should be persuasive because Job's salvation was at stake.

Prior to Darwin, it was common practice to explain scientific evidence in the light of the bible — what people knew to be true and to introduce ad hoc factors to assert a consistency with the bible. On page 202 of *Origin of Species*, Darwin writes: "...I would almost as soon believe with the old and ignorant cosmologists, that fossil shells had never lived, but had been created in stone so as to mock the shells now living on the seashore." These old and ignorant cosmologists were creationists defending God's word against the natural evidence (and setting up creationism for a fall) because creationism was *important*. These creationists were polarising Darwin by adding to his confidence. Darwin could see that there was something wrong with the arguments of creationists, just as Job's friends could see that there was something wrong with Job's arguments.

Darwin was concerned with the cause of overthrowing the idea of 'separate creations' as he wrote in chapter 2 of *Descent of Man:* "Some of those who admit the principle of evolution, but reject natural selection, seem to forget, when criticising my book, that I had the above two objects in view; hence if I have erred in giving to natural selection great power, which I am very far from admitting, or in having exaggerated its power, which is in itself probable, I have at least, as I hope, done good service in aiding to overthrow the dogma of separate creations." And when referring to Lamarck, on page 54 of his *Origin of Species*, Darwin said: "He first did the eminent service of arousing attention to the probability of all change in the organic, as well as in the inorganic world, being the result of law, and not of miraculous interposition."

If there are issues of high importance (such as safety issues), then concerned citizens will have a cause. There are no excuses but we should not be surprised to find bias in creationist's arguments because creationism is very important to the creationist. We expect something better from scientists — we expect scientist to be equally critical of the arguments of other scientists, evolutionists, and creationists. We expect errors to the left and an equal number of errors to the right because we suppose science to be objective and honest and not to care about the outcome. We expect science to preach the weakness of scientific theories from the roof tops. We trust that if we do not understand all the details then we need not worry since the matter is well taken care of by lots of people more qualified than we are and who are only interested in finding the truth. However evolutionary science is different to ordinary science not just because of philosophical naturalism but also because the theory of evolution is important science.

Important science is different from ordinary science. A scientist can be an environmentalist, for example, but when he is an environmentalist then he is a scientists with an objective or a purpose. A scientist with a purpose will never flaunt problems with their truth because the 'truth for truths sake' principle of science is replaced by the objective and then we will only hear how there are no problems with their truth. Thus, an objective and the fact that we live in an imperfect world causes us to ignore the process of truth. An anti-creationist and a creationist might object to this claim, but this is more of the same — it is a response that comes from an inability to separate truth from the-truth.

Science education in our schools is perhaps the biggest single factor that drives the importance of evolution. Because of the need for separation of religion and state, science education in public schools is a legal issue. In America, the focus is on the fact that the first amendment to American Constitutional requires that 'the Federal Government shall not establish or disestablish a religion' with emphasis on what should be included in the government school curriculum. It is said that creationists have used intelligent design to attempt to step around such laws. For example, the NCSE article *Intelligent Design Not Accepted by Most Scientists* argues that: "ID proponents may argue that a neutral-sounding "intelligence" is responsible for design, but it is clear from the "cultural renewal" aspect of ID that a deity — in particular, God as He is conceived of by certain conservative Christians — is envisioned as the agent of design. While schools can take no position on this view as religion, it cannot be regarded as science." And another National Centre for Science Education article titled *What is Intelligent Design Creationism?* argues that: "IDC proponents usually avoid explicit references to God, attempting to present a veneer of secular scientific inquiry. IDC proponents introduced some new phrases into anti-evolution rhetoric, such as "irreducible complexity" (Michael Behe: *Darwin's Black Box*, 1996) and "specified complexity" (William Dembski: *The Design Inference*, 1998), but the basic principles behind these phrases have long histories in creationist attacks on evolution. Underlying both of these concepts, and foundational to IDC itself, is an early 19th century British theological view, the "argument from design."

Thinking like NCSE, we should be able to say that there should be no doubt that because intelligent design theory supports the Biblical interpretations, then of course intelligent design theory is just an another creationists agenda — another attempt to get some narrow religious viewpoint into the educational system. Because Intelligent Design does not say 'God' but 'God' is the obvious interpretation, it is said that the reasons for the use of the 'intelligent design' terminology is purely agenda. Intelligent design theory does not mention God or religion, but does imply God to many of us. According to p25 of the National Academy of Sciences booklet, "No body of beliefs that that has its origin in doctrinal material rather than scientific observation, interpretation, and experimentation should be admissible as science in any science course. Incorporating the teaching of such doctrines into a science curriculum compromises the objectives of public education." In other words, we no longer need to consider whether an argument is credible because all we need to do is see if it is inspired by the Bible. Using similar thinking, American courts have determined that the critical issue is whether or not intelligent design theory has religions motivations.

It makes some sense to say that the theory that all life was created in six days is religion not science because we want the reliability of science to be based on observation and experiment. We want science to be knowledge rather than revelation. And it is true that if the American courts could say that intelligent design *only* has religious motivations or that the only support for creationism was in the Bible, then the religion not science statement begins to make some sense. However, when the issues are simple, the correct way of determining whether some doctrine *only* has religions motivations is to determine whether it has any ordinary credibility (independent of the Bible and religion).

It makes some sense to assume that the American Constitution was intended to at least keep things like the Catholic Church and the Church of England out of America. I doubt that the constitution was also intended to evict Jesus from the American Government, so as not to upset Darwin, Buddha, Mohammed and lots of cows in India. However, regarding the issue of whether intelligent design should or should not be included in the education system, the argument is much simpler and we can proceed on the worst case basis. This is because the American Constitution is really about prejudgement thinking to give us the best guess when the issues become too complex for the uniformed bystander.

It would not make sense to say that 'the American Constitution should not establish or disestablish logic, mathematics, science, etc', but it would make some sense in the interest of fairness to say this of a particular denomination or religion. This makes sense because the thing about religion is that there are a lot of different opinions about which religion or denomination is true. It would also make sense that the Australian Government should not establish or disestablish history, but this does not mean that history cannot be a part of the public school curriculum — but only that the history should not be biased. Some years ago, John Howard mentioned the need to teach more Australian history in our classrooms and he then took time to clarify that he did not mean that he wanted a particular 'Australian' version (such as the black version or the white version) of history taught. The 'establish or disestablish' requirement is something like a need to eliminate all bias on the part of the Government toward or against a particular denomination/religion. This bias could be in the form of information and education or it could be in the form of government funding.

The real issue addressed by religious neutrality is complexity. Such issues arise because the bystander is not easily able to determine if denomination A or denomination B is speaking the truth and therefore, he decides to take a neutral position.

Although the courts may be able to legitimately use weak prejudgements to sort out religious issues, this is only when issues of religious truth are too complex for the courts and uninformed bystanders to sort out. Prejudgements are what we use to give us a best guess when we cannot do any better.

The argument for creationism (but not Biblical creationism), is very simple and direct and therefore the courts should easily be able to determine credibility directly by considering the evidence and arguments. We all think in terms of a designer (or designers or creators if we want to be general) given the need to explain a machine because this is ordinary thinking that can be argued independent of the Bible. Biochemical machines are much more complicated than our mechanical machines, so we might need some sort of super-creator (God) to explain biochemical machines. What this means is that, if intelligent design theory has religious implications then this is not the fault of either the intelligent design argument, but it is the fault of ordinary reliable thinking, that comes from our understanding of complexity and the best explanation reasoning. The explanation of design complexity by reference to an intelligent designer is amongst the most reliable forms of reasoning available to man. If we found a machine in some very old rocks we would conclude a designer regardless of how scientifically certain we were that there were no designers around to make the machine.

Intelligent design theory is ordinary (non-religious), exceptionally reliable thinking with religious implications and to rule against 'ordinary reliable thinking that establishes ideas with religious implications' is to establish of Holbach's vile hatred for God. The obvious solution to the problem is intelligent design, and it is only because anti-creationists have confused themselves with their own rhetoric that they see the need to define science so that intelligent design is not a solution. We should say that intelligent design is not established by the government but it establishes itself. Or rather, regardless of whether we are a creationist, anti-creationist, atheist, agnostic, Hindu or Buddhist, (but providing we do not have an unbeatable confidence in naturalism or a hatred for the idea of God) we should all agree that intelligent design theory comes from EXTREMELY reliable thinking.

Spokesmen for evolution like to ignore the problem of the first life form and the issue of evolution compatible complexity and remind readers that biochemical machines are different from manmade machines because they can reproduce.

Biochemical machines can reproduce and being able to reproduce means that we can suppose that they evolve providing we believe in an evolution compatible understanding of complexity and providing we can explain the first life form that could reproduce. The fact that biochemical machines can reproduce should mean that we all understand that biochemical machines have a kind of complexity that we do not begin to understand.

Philosophical naturalism denies the existence of a real God that can do things, because science says that super-duper naturalism has done everything. Philosophical naturalism insists that the evidence for a creator must be explained away. Philosophical naturalism is the God-hating religion established by all western governments. They are the facts and it would be good if our spokesmen for science could come to terms with the facts, rather than using the confidence of philosophical naturalism and the importance of the theory of evolution to trick themselves into more confusing arguments.

Given the need to maintain their independence from popular opinion and political pressures, the American courts, like Australian courts, should be good at standing up to the pressures of science. However, the American courts are also given over to philosophical naturalism because Americans are westerners. And it may be that American courts do not understand that there is such a thing as the pressure of science.

American courts perhaps think that science has the answers to these questions and that science has no reason to be biased, whereas we all understand that fundamentalists have every reason to be biased. If science had the answer to these questions then science would not be invoking infinite universes and clay crystal theories. And if science was so different to fundamentalism, then it would not be denying the influence of philosophical naturalism.

Concerned people want all the errors in one direction not because they have an evil agenda to deceive people but because truth can be dangerous. A cause (or a best interest) means a need to push something in front of truth, because we do not live in a perfect world and our truth is important. Cairns-Smith, for example, is convinced that there must be a natural solution to the creation of life, but because he is not concerned he argues on pages 4 and 39 of his book *Seven Clues to the Origin of Life:*

> The optimism persists in many elementary textbooks. There is even, sometimes, a certain boredom with the question; as if it was now merely difficult because of an obscurity of view, a difficulty of knowing now the details of distant historical events. What a pity if the problem had really become like that! Fortunately it hasn't. It remains

a singular case (Sherlock Holmes' favourite kind): far from being a million ways in detail in which evolution could have got under way, there seems now to have been no obvious way at all. The singular feature is in the gap between the simplest conceivable version of organisms as we know them, and components that the Earth might reasonably have been able to generate. This gap can be seen more clearly now. It is enormous.

What is wrong with the story that I have been telling is that it hardly touches the real difficulties: the difficulties that I was piling up over the first four chapters.

I will grant that the path of chemical evolution seems sensible and in the right direction. There are the obvious puddles to be avoided and some of the flagstones are a bit uneven, perhaps, but there is the promise of an easy walk up to the foothills of the mountain that we can see straight ahead of us. It is a promise that is unfulfilled. The trouble with this path is that it leads toward but does not lead us to expect a sudden near-vertical cliff-face. Suddenly in our thinking we are faced with the seemingly unequivocal need for a fully working machine of incredible complexity: a machine that has to be complex, it seems, not just to work well but to work at all. Is there cause to complain about this official tourist route to the mountain? Is it just a garden path that we have been led along — easy walking but never getting anywhere? I think it is...

Even if amino acids (for example) had been in a 'probiotic soup', what use would they have been, long before their key use now (to make protein) had been hit on? *It is the whole machine that makes sense of its components.*

Point two. Subsystems are highly INTERLOCKED within the universal system. For example, proteins

> are needed to make catalysts, yet catalysts are needed to make proteins. Nucleic acids are needed to make proteins, yet proteins are needed to make nucleic acids. Proteins and lipids are needed to make membranes, yet membranes are needed to provide protection for all the chemical processes going on in a cell. It goes on and on. The manufacturing procedures for key small molecules are highly interdependent: again and again *this* has to be made before *that can* be made — but that had to be there already. The whole is presupposed by all the parts. The interlocking is tight and critical. *At the centre everything depends on everything.*

As noted at the beginning of section 6.1, Cairns-Smith and Shapiro are open about their philosophical naturalism. We could say that they are not concerned about winning the argument against creationism. Cairns-Smith's book is a discovery book with a discovery attitude. Although Cairns-Smith is a committed evolutionist who is sold out to philosophical naturalism and the objective of his book is to present a natural means by which the first life form could have been created, Cairns-Smith's book could be part of Sunday School curriculum — assuming the Sunday School participants had a reasonable understanding of philosophical naturalism. To understand both Cairns-Smith's book and philosophical naturalism is to understand that intelligent design theory is common sense.

Given an understanding of philosophical naturalism, Dawkins' books and the Origin of Species are also an inspiration for creationism. Not to the same degree as Cairns-Smith's book, because Cairns-Smith is not out to persuade. On page 6 of *Science and Creationism* NAS argue:

> For those who are studying the origin of life, the question is no longer whether life could have originated by chemical processes involving non-biological components. The

> question instead has become which of the many pathways might have been followed to produce the first cells.

May all the courts agree that unlike Cairns-Smith's book, the NAS booklet should never be a part of any school curriculum. Ordinary truth says that Cairns-Smith's evolving clay crystals is one of the more celebrated solutions, whereas scientific truth says that the self-assembly of a biological machine is not such a difficult problem after all. Infinite universes and clay crystals being only two of the possible solutions. Let all the courts understand that western science has an interest in promoting scientific truth over ordinary truth.

Until science fixes the natural bias that comes from philosophical naturalism and the importance of evolution, American courts may need to think for themselves. However, there may be issues when American courts will have no choice but to listen to the experts. Spokesmen for evolution could argue that there are other reasons that evolutionists believe in evolution that I have not discussed — topics that may be difficult the courts to get a handle on. The fossil record is one such issue of truth that speaks about the theory of evolution. From p 28 of *Science and Creationism* the NAS experts tell us:

> Scientists most often use the word "fact" to describe an observation. But scientists can also use fact to mean something that has been tested or observed so many times that there is no longer a compelling reason to keep testing or looking for examples. The occurrence of evolution in this sense is a fact. Scientists no longer question whether descent with modification occurred because the evidence supporting the idea is so strong.

It is a fact that there were once dinosaurs and other life forms very different from what is around us today. If the fossil evidence supports descent with modification, then we might be able to say 'yes an evolutionary process seems impossible yet we have evidence to suggest that some kind of evolution (descent with modification) happened.' If it truly is the case that the fossil record makes such a clear statement about evolution, then how are we going to find out? Unless we have lots of spare time on our hands, then we will need to rely on the opinion of the expert. From p292 of *Origin of Species*, one expert asks:

> Why then is not every geological formation and every stratum full of such intermediate links? Geology assuredly does not reveal any such finely graduated organic chain; and this, perhaps, is the most obvious and gravest objection which can be urged against my theory. The explanation lies, I believe, in the extreme imperfection of the geological record.

In an intellectual climate when evolution was not the status quo, Darwin saw the geological record as being a problem for his theory rather than support for his theory. Evolutionist, palaeontologist, and Harvard professor Steven Gould once asserted in an intellectual climate that supported evolution that the absence of transitional forms is the trade secret of palaeontologists. On pages 153 to 154 of *The God Delusion,* Dawkins cites J. Haldine as his expert:

> It is utterly illogical to demand complete documentation of every step of any narrative, whether in evolution or any other science. You might as well demand, before convicting some body of murder, a complete cinematic record of the murderer's every step leading up to the crime, with no missing frames. Only a tiny fraction of corpses fossilize, and we are lucky to have as many intermediate fossils as we do. We could easily have had no fossils at all, and still the evidence for evolution from other sources, such as molecular genetics and geographical distribution, would be overwhelmingly strong. On the other hand, evolution

> makes the strong prediction that if a *single* fossil turned up in the *wrong* geological stratum, the theory would be blown out of the water. When challenged by a zealous Popperian to say how evolution could ever be falsified, J. B. S. Haldane famously growled: 'Fossil rabbits in the Precambrian.' No such anachronistic fossils have ever been authentically found, despite discredited creationist legends of human skulls in the Coal Measures and human footprints interspersed with dinosaurs'.

It is nice to know that the fossil strata is so uniform. However, I am still in need of an expert to tell me what that means? What if the pattern was different; would that be a contradiction of the theory of evolution or just a contradiction of the known pattern? Is the pattern truly predicted by the theory of evolution, or it is explained by the theory of evolution because once we have discovered a pattern, then the pattern will make predictions independently of the theory?

What sort of expert do we want to answer such questions? Let all the courts agree that we do not want the expert who argues that naturalism has never been and is not a part of science. And we don't want the expert that tells us that evolution is typical science. And we do not want the expert who tells us 'that there are just so many ways that life could come to be that we just don't know which one to pick?' In other words, we do not want experts who are concerned scientists. We want an expert that has an interest in telling the rest of us about the weakness and problems with the evidence so that we do not need to find a semi-informed creationists to ask the appropriate questions?

An article titled *The Church of Darwin* by Phillip Johnson suggests that we should get our experts from China: "A Chinese paleontologist lectures around the world saying that recent fossil finds in his country are inconsistent with the Darwinian theory of evolution. His reason: The major animal groups appear abruptly in the rocks over a relatively short time, rather than evolving gradually from a common ancestor as Darwin's theory predicts. When this conclusion upsets American scientists, he wryly comments: "In China we can criticize Darwin but not the government. In America you can criticize the government but not Darwin."" This is perhaps because at this point in time, Chinese scientists are not concerned citizens and philosophical naturalism is predominantly a western disease.

Most creationists are Christians. If we say that spokesmen for evolution are concerned, then we must say that most creationists have a heaven and hell concern. Self-interest helps me with this problem. I do not want to be ashamed so I take some care. And it helps to recognise that even though there is a real hell, God did not stop the Roman soldiers, who were guarding the body of Jesus, from telling people that the disciples had stolen the body, despite the bodily resurrection of Jesus being the single most important truth of the Bible. God chose not to give an explanation of the fossils in the Bible and thereby allowed science to devise what many people consider to be a reliable argument against a real God who can do things. Even though hell will freeze over before I take a critical attitude to the Gospel, it helps me to understand that God considered bringing a judgement on Job's friends who were speaking for God and it helps me to understand that philosophical naturalism was in some sense perpetrated by the fact that God hides himself.

Scientists are concerned citizens. Should science argue like Cairns-Smith or like the *National Academy of Sciences*? Does science want to argue for the best interest or to argue for whatever may come? Should science support good arguments that are against the best interest of science? Do scientists want their arguments focused on the ends or the means? Suppose a scientist discovered some facts that supported 'creationism' would the *National Academy of Sciences* want to disclose the facts or would they want to wait until someone found a satisfactory explanation for the facts?

The fact that the Epicurean spleen has too much understanding of an evil creationist agenda, will mean that the Epicurean spleen will not be able to consider that science would ever conceal the facts of science. However I am suggesting that NAS might want to conceal the facts because NAS thinks the facts would be truly misleading and because they think that evolution is important. In other words I am asking whether science would be willing to subvert science by being open about the facts that speak against the theory of evolution? Given Holbach's confidence in philosophical naturalism, all facts that suggest a real creator are bound to be interpreted as misleading facts.

Evil agenda thinking, bitterness and contention are all spiritual problems. If we suppose that the Epicurean freethinkers came into the understanding that they were mistaken then they would not be able to transfer their judgements about creationists to themselves because there is a spiritual aspect to their kind of freethinking. We also know that there is a spiritual aspect to many arguments because we would all be much better at finding ourselves wrong if our life depended on the truth of our argument. And besides all these issues, naturalism is a religious theory about who we are and where we come from. Because there is a spiritual aspect to these issues, I would like to put on my preachers hat and finish up by throwing some Scriptures at these issues.

Contention guarantees a need to win the argument, point-scoring, derogatory arguments, missing the real issues, never conceding an error, correcting irrelevant errors, creating smoke screens and a failure to properly and clearly address arguments point-by-point, etc. In other-words, overcoming the opposition and winning an argument that we shouldn't win. Christians suffer from the same spiritual problems. If the Holy Spirit is the spirit of truth then the things like bitterness and contention, that James (and Paul in his letter to the Corinthians) addresses could only work against the anointing of the Holy Spirit in God's ministers. James 3:17 says that wisdom should first be pure then peaceable, gentle and easily entreated and not derogatory. Proverbs 11:1 says: "A false balance is abomination to the LORD: but a just weight is his delight". We back our ideas and give ourselves reason to contend with anyone who would challenge our ideas because when somebody criticises our idea then they criticise us. But, there is another more sinister enemy behind these things, namely the perceived 'deceptive practices' and the 'agendas' of our enemies. Deconstructing our evil agenda thinking is to beat a few demons around the head.

Popper saw the solution to the problem of truth in terms of critical thinking. Critical thinking was a natural approach for Popper because it is similar in attitude to trying to falsify a theory and it is opposite to trying to confirm that a theory is true. Popper p *50 of Logic of Scientific Discovery:*

> A system such as classical mechanics may be 'scientific' to any degree you like; but those who uphold it dogmatically — believing, perhaps, that it is their business to defend such a successful system against criticism as long as it is not *conclusively disproved* — are adopting the very reverse of that critical attitude which in my view is the proper one for the scientist.

NCSE have embraced the commission of 'defending the teaching of evolution and climate science' and this is perhaps explains how it came to be that NCSE argues that philosophical naturalism is irrelevant to science. If the claim that science is based on philosophical naturalism seems to undermine the truth of the theory of evolution and we are stuck on defending evolution, then like every good fundamentalist we will see the need to speak wickedly and deceitfully for our god.

Thinking is a kind of looking. Critical thinking is a matter of looking in all directions, which takes effort and is inherently inefficient. Critical thinking is a matter of time, effort and attitude and it should be a good thing. Unfortunately, though labels like the the critical thinking label can give us a wrong understanding of who we are.

The enthusiastic Professor Schafersman in his web document *Naturalism is today and Essential Part of Science* gives a definition of critical thinking in the two or three paragraphs immediately under his sub-section: What is Science? While the professor may have the intellectual skills associated with good thinking, he manages to make his critical thinking a part of the problem.

Professor Schafersman's webpage preaches Dawkins' more logical form of philosophical naturalism. Like Holbach, he is open and honest about where his confidence comes from.

Instead of critical thinking, Holbach esteemed courageous enlightened thinking. In part 1, chapter 1 of his *System of Nature,* Holbach wrote of the *enlightened man*: "The enlightened man is man in his maturity, in his perfection; who is capable of advancing his own felicity, because he has learned to examine, to think for himself, and not to take that for truth upon the authority of others, which experience has taught him a critical disquisition will frequently prove erroneous." And that would be Holbach of course. Immanuel Kant said something very similar. In part 2 chapter V, Holbach writes of Newton's religious thinking:

> Let us now see, if the immortal Newton, the great luminary of science, the champion of astronomical truth, will afford us clearer notions, more distinct ideas, more certain evidence of the existence of immaterial

substances. This great man, whose comprehensive genius unravelled nature, whose capacious mind developed her laws, seems to have bewildered himself, the instant he lost sight of them. A slave to the prejudices of his infancy, he had not the courage to hold the lamp of his own enlightened understanding to the agent theology has so gratuitously associated with nature; he has not been able to allow that her own peculiar powers were adequate to the production of that beautiful phenomena, he has with such masterly talents so luminously explained. In short, the sublime Newton himself becomes an infant when he quits physics, when he lays aside demonstration, to lose himself in the devious sinuosities, in the inextricable labyrinths, in the delusive regions of theology. This is the manner in which he speaks of the Divinity: "This God," says he, "governs all, not as the soul of the world, but as the lord and sovereign of all things...

The necessary existence of the Divinity is precisely the thing in question; it is this existence that it was needful to have verified by proofs as clear, by evidence as distinct, by demonstration as strong, as gravitation and attraction. One would have hardly thought it possible the expansive capabilities of Newton would not have compassed it. But oh, unrivalled genius! so mighty, so powerful, so colossal, while yet you was a geometrician; so insignificant, so weak, so inconsistent; when you became a theologian; that is to say, when you reasoned upon that which can neither be calculated, nor submitted to experience; how could you think of speaking to us on a subject which, by your own confession is to you just what a picture is to a man born blind? Wherefore quit nature, which had already explained to you so much? Why seek in imaginary spaces those causes, those powers, that energy, which she would have distinctly pointed out to you, had you been willing to have consulted her with your usual

> sagacity? The gigantic, the intelligent Newton, suffers himself to be hoodwinked--to be blinded by prejudice; he has not courage to look a question fairly in the face, when that question involves notions which habit has rendered sacred to him; he turns his eyes from truth, he casts behind him his experience, he lulls to sleep his reason, when it becomes necessary to probe opinions full of contradictions, yet fraught with the best interests of humanity. Let us, however, continue to examine how far the most transcendent genius is capable of leading himself astray, when once he abandons experience, when once he chains up his reason, when once he suffers himself to be guided by his imagination.

Holbach convinces himself that philosophical naturalism is enlightened knowledge, which only comes to those who have sufficient courage and maturity, which is to say that Holbach understands the weakness and failings of people like Newton, that he is not surprised at what such people say.

On page 152 of *The God Delusion*, Dawkins says of scientific thinking: "It is an essential part of the scientific enterprise to admit ignorance, even to exult in ignorance as a challenge to future conquests." Again there is a spiritual aspect to statements like that because although that might be true for most science, it is not true for the theory of evolution. Dawkins has succeeded in misunderstanding who he and his fellow scientists are. On page 320 of *The God Delusion* Dawkins writes:

> Maybe scientists are fundamentalist when it comes to defining in some abstract way what is meant by 'truth'. But so is everybody else. I am no more fundamentalist when I say evolution is true than when I say it is true that New Zealand is in the southern hemisphere. We believe in evolution because the evidence supports it, and we would abandon it overnight if new evidence arose to disprove

> it. No real fundamentalist would ever say anything like that. It is all too easy to confuse fundamentalism with passion. I may well appear passionate when I defend evolution against a fundamentalist creationist, but this is not because of a rival fundamentalism of my own. It is because the evidence for evolution is overwhelmingly strong and I am passionately distressed that my opponent can't see it - or, more usually, refuses to look at it because it contradicts his holy book. My passion is increased when I think about how much the poor fundamentalists, and those whom they influence, are *missing*. The truths of evolution, along with many other scientific truths, are so engrossingly fascinating and beautiful; how truly tragic to die having missed out on all that! Of course that makes me passionate. How could it not? But my belief in evolution is not fundamentalism, and it is not faith, because I know what it would take to change my mind, and I would gladly do so if the necessary evidence were forthcoming.

Although Dawkins' arguments do seem open and honest, in the interest of not being confused about who he is, it would be better if Dawkins understood that his openness and honesty was not his fault. Like Dawkins I believe that my truth will be manifest to all who will open their eyes. However, a discovery of another Piltdown man might send me into a *NAS* spin. Even without another Piltdown man things could be different for me if I did not understand philosophical naturalism. In other words when it comes to the topic of creationism, religion friendly spokesmen for evolution live in an imperfect world whereas Dawkins and I currently live in a near perfect world. I am trying to change that for Dawkins.

Dawkins' 'gladness' seems to have lost track of the fact that he is concerned about creationism. Dawkins' 'gladness' doesn't seem to understand that if a creationist were to show him a gaping hole in one of his arguments, then he would have another problem with truth aside from the fact that he is a concerned citizen. And despite the fact that Holbach was open and honest, Holbach might not be so glad about understanding that his extreme confidence came from a delusion. In the forward of David Mills' book *Atheist Universe*, Dorion Sagan writes:

> Creationists are funny. They want to be taken seriously as scientific and have their or their comrades' writings taught as science to our children in schools. But their attitude is unscientific. Rather than engaging in open-minded investigation to figure out how things are done, at the first glimmer of mystery they throw up their hands and say it is beyond science. This is like not knowing how a magic trick is done and thinking no one else can know, either. We can give up trying to understand because God-who in their view is like a 2000-year-old petty Middle Eastern tyrant, quick to anger and condemn to the eternal prison of Hell those who don't obey Him - must have done it. As soon as they (*to sound* scientific) ascribe the phenomenon to irreducible complexity, Intelligent Design, the Flying Spaghetti Monster, or something else, they have abandoned the search. Needless to say, suggesting something is beyond science evinces neither the spirit of entrepreneurial inquiry and American ingenuity nor is it, in any way, shape or form, science. Science does not stop at some artificial limit of inquiry. It continues on. It, to borrow an apposite religious term, perseveres. Creation "science" and Intelligent Design theory do not persevere. They throw in the towel.

Science, in the words of quantum physicist David Bohm, is about finding the truth *whether we like it or not.* Apparently some fundamentalists are not comfortable with the truth-so much so that they have gone undercover,

> pretending to be scientists, except that (as David Mills shows here) they invent facts and contradict themselves to arrive at their preconceived conclusions.

Science has no problem whatsoever that science has been stopped in the God direction and understanding that scientists are different to creationists might not be helpful to science. Not if we want to avoid another 150 years of the same.

Inappropriate generalisations continue to be a large part of the problem. When we do not discriminate between 'how things are done' and 'how things were created' as Dawkins and Sagan do, then we are in danger of being tricked by the pattern. And when we imagine that science, that is not tested by technology, is like science that is tested by technology, then we are in danger of tricked by the pattern.

Science has been extremely successful, because most scientific theories are subject to repeatability and because there are lots of scientists focusing on the problems of science and sharing their discoveries with fellow scientists. Evolution cannot be justified by use of the principle of repeatability and therefore the theory of evolution is different to science that is tested by technology.

Science objects to the idea of singling out evolution for special treatment. Science argues that the special feature of creationists arguments is the 'religious sensitivities' or the religious inspiration of the argument. Like other 'God of the gaps arguments', this argument effectively tells us that the reason we have creationism is because we have religion, which is to say that creationism made reference to a creator only for the purposes of having a religion, rather than because we need a creator to explain creation.

The theory of evolution is special in many ways. Evolution is stuck on philosophical naturalism. Evolution is important science. Evolution is a historical theory. The problem of creation is special because a logical feedback scheme is the only possible solution because that there cannot be a creation law of the universe. Evolution is one of the few theories of science where the philosophy that 'we will accept a supernatural solution only if a natural solution does not seem possible' is unacceptable bias because evolution is concerned with creation rather than operation-and-maintenance. Even the 'excrement of ants' creation myth (from p 316 of Dawkins *Blind Watchmaker*) is better thinking than modern science because Dawkins' primitive Africans recognised creation as a special problem of science.

Evolution is special because it beckons the allegiance of science. Evolution is special because when God comes out of hiding and the events of the book of Revelation come to pass, science will abandon the theory of evolution, but it will hang on to physics and chemistry.

Science is generally open and transparent. Science is generally only looking for the truth and does not care about the outcome. But if we want help scientists understand who they are then we need some sub-divisions and some understanding. Science that is not concerned is open and transparent. Science that is concerned is not. Science that concerns operation-and-maintenance laws may be free from bias. Science that concerns creation is subject to the gravity of philosophical naturalism.

The Discovery Institute has argued intelligent design but science has responded with words and definitions like 'intelligent design is not science' because science is not allowed to say things like 'it seems like a creator created'. Science is very good at defending itself from creationist arguments but incapable of expressing a creationist argument.

Behe tried to sharpen our understanding of complexity but nevertheless on page 87 Hitchens said: "Based on the reconstruction of ancient genes from extinct animals, the researchers were able to show how the nontheory of "irreducible complexity" is a joke." It would seem that 150 years after Darwin and science does not understand that the theory of evolution requires that there is such a thing as evolution-compatible complexity, because it is impossible for an enhancement algorithm (i.e. a one best step at a time algorithm) to create ordinary complexity.

Science insists that the 'the problem of the first life form is in the bag' apparently because there are other theories in competition with the clay crystal theory and the infinite universe theory again because science knows that philosophical naturalism is the truth and because science must never say that 'the first life form problem seems to need a God solution'.

Why should science need the help of Professor P. Johnson? Why can't science work out the relevance of philosophical naturalism for itself? Despite Professor P. Johnson's efforts, science seems to be happy to convince itself that science is not influenced by philosophical naturalism.

If we claim that it is natural that science is concerned about evolution, then we should agree that it is not so natural that science does not seem to be aware of the bias of science that comes from philosophical naturalism or the bias that comes from the importance of the theory of evolution. It should be that after 150 years that science has some understanding of our scientific culture.

We creationists may be fundamentalists but our religion is based on an open faith whereas philosophical naturalism is based on a delusion. Sometimes a concealed delusion.

Scientists sometimes argue that science cannot accept the magic of supernaturalism otherwise there would be no point experimenting. Does this mean if a scientist was dragged along to a meeting with one of Smith Wigglesworth's disciples and supernaturally healed he would not be able to do science experiments the following day? If science has no interest in correcting this argument, or the 'creationism is not science argument'etc; then something needs to be fixed.

It could be argued that it is natural for scientists to think like scientists. Even we creationist think like scientists. What we cannot accept is that scientists do not know that they think like scientists. My real complaint is the same complaint raised in the conclusion to chapter 4. I am not so much thinking that 150 years have gone by and science has not addressed the natural bias of science, as 150 years have gone by and science has no understanding of the natural bias against creationism. When the topic is the theory of evolution, rather than science being about finding the truth whether *we like it or not* science has no interest in any truth that does not support the theory of evolution.

We fundamentalists know that we are fundamentalists. Science understands itself to be open-minded, freethinking and truth seeking. Proverbs 26:12 says: "Seest thou a man wise in his own conceit? [there is] more hope for a fool than of him." The self-flattery of science like the delusion of philosophical naturalism is a generalisation error. The fantastic achievements of science are what make science vulnerable to this generalisation error.

According to Proverbs 20:6: "Most men will proclaim every one his own goodness [critical thinking is a type of goodness]: but a faithful ['emuwn] man who can find?" This is a two part proverb suggesting that a self-professed freethinker is a man who is not faithful and that he does not know who he is. We need to be thinking of something subtle here because it is applicable to most men. Perhaps it would be better to speak in terms of 'living in our dreams' rather than 'self-flattery', because we all like to live in our dreams.

The Hebrew word for 'faithful' is first used in Deuteronomy 32:20: "And he said , I will hide my face from them, I will see what their end [shall be]: for they [are] a very froward [tahpukah] generation , children in whom [is] no faith ['emuwn]". This Deuteronomy Scripture suggests that the correct translation may be a word that is opposite to the froward spirit. Proverbs 16:28 uses the word 'froward' in terms of disloyal speech and backbiting: "A froward [tahpukah] man soweth strife: and a whisperer separateth chief friends." Proverbs 14:5 says: "A faithful ['emuwn] witness will not lie: but a false witness will utter lies".

It seems that a practical approach to truth cannot work on its own. For example, if we were to argue that, 'for every ten people that hear a "let's do God a favour" argument there may be eventually lots more who are turned off creationism', then we are effectively saying that there is a 'purpose to to be served by telling the truth' when 'purpose' is the reason why we have a problem with truth. Fundamentally truth does not care about purpose whereas deception is all about purpose.

Instead of coming up with reasons why we should tell the truth, we could try the other extreme and focus on truth as a principle. Psalm 15 argues for putting principles before profit:

> [[A Psalm of David.]] LORD, who shall abide in thy tabernacle? who shall dwell in thy holy hill? He that walketh uprightly, and worketh righteousness, and speaketh the truth in his heart.
>
> [doesn't deceive himself]
>
> [He that] backbiteth not with his tongue, nor doeth evil to his neighbour, nor taketh up a reproach against his neighbour. [no bitterness]
>
> In whose eyes a vile person is contemned; but he honoureth them that fear the LORD. [He that] sweareth

> to [his own] hurt, and changeth not. [puts principle before profit.]
>
> [He that] putteth not out his money to usury, nor taketh reward against the innocent. He that doeth these [things] shall never be moved. [a nice guarantee that is irrelevant to our topic. If we are not in relationship with God then we need to be moved.]

Putting principles before profit (whether it be social profit or monetary profit or any other kind of profit) is similar to arguing for truth as a principle. Arguing for truth as a principle is difficult because principles are those things for which there is little or no practical justification. Because the self-professed man of principles is a man who does not know himself, we can say that in our dreams we all have principles, because in our dreams principles reward us, but in real life principles hurt us. A business man who wants to be effective will very often tell the truth, consider others and in general do the right thing both in his dreams and in real life, because doing the right thing rewards us. Unless we belong to a narrow class of people (anti-social criminals and Jesus), then we all need the acceptance of society for our comfort and survival. The effective man will take advantage of principles and lie when it suits. Or as Jean Giradoux said: "the secret to success is sincerity — once you can fake that, you've got it made". The effective man is the prudent man who wants to be seen to have principles. The ineffective man is the man of Psalm 15 that "sweareth to *his own* hurt and changeth not." However, as noted above, an emphasis on things like principles can cause us to misunderstand who we are.

Psalm 15 is not truly an argument for principles because it the last line includes the promise: "He that doeth these [things] shall never be moved." And although Popper advocated critical thinking, according to the last words of a long footnote on p 258 of *The Logic of Scientific Discovery*,

Popper's complete solution to this problem would be for every spokesmen for evolution, to develop the strongest possible argument demonstrating why creationism is the truth. This is a personal strategy for individual scientists who understand that when it comes to problems like evolution we do our best critical thinking in our dreams. According to Popper, the thinking we want that is much better than freethinking or scientific thinking is really a willingness to find ourselves wrong. Popper's solution is more about the fact that individuals deceive themselves before they deceive their audience.

Popper's solution could be applied to each individual issue of truth that speaks against the theory of evolution. Scientists who like to think with generalisations could start from the understanding that evolutionary science has been established 'by the same means as the NAS statement about the first life form', or 'by the same means is the argument for methodological naturalism' etc. The concept of principles then becomes a matter of how we allow issues of truth to speak independently of the-truth. The concerned citizen might choose to only allow such issues to speak to himself. But an unconcerned scientists should give a loud voice to each issue of truth that opposes accepted science. Dawkins says that he would gladly change his mind but we need scientists like Dawkins to give a voice to these issues of truth that speak against the theory of evolution — regardless that they believe that such issues may mislead people away from 'the-truth' of philosophical naturalism.

When it comes to the theory of evolution western science does not have any trouble makers apart from a few creationists. Why is there even a need to say such things? Science does not have any trouble makers because evolution is one of a few ideas that beckons the allegiance of science against the understanding that there is a real God. Science needs some trouble makers as a way of making fellow scientists ashamed of wrong arguments that end up at the right place.

If we are going to insist that there are so many ways that the fist life form could have come about then someone needs to be ashamed. If we are going to insist that philosophical naturalism does not influence science when blind Freddy can see that it does then someone needs to be ashamed.

As mentioned previously, there is a spiritual aspect to philosophical naturalism. Biblical creationists are stuck on the idea that God is the creator and scientists are stuck on super-duper naturalism. Holbach argues that the truth of philosophical naturalism will be obvious to all those who would have the courage to open their eyes to his world view but Romans 1:17-32 tells us that we are without excuse for not believing in a creator, because the evidence of creation is all around us:

> For the wrath of God is revealed from heaven against all ungodliness and unrighteousness of men, who hold [or hold down or suppress] the truth in unrighteousness;
>
> Because that which may be known of God is manifest in them; for God hath shewed [it] unto them.
>
> For the invisible things of him from the creation of the world are clearly seen, being understood by the things that are made, [even] his eternal power and Godhead; so that they are without excuse:
>
> Because that, when they knew God, they glorified [him] not as God, neither were thankful; but became vain in their imaginations, and their foolish heart was darkened.
>
> Professing themselves to be wise, they became fools,
>
> And changed the glory of the uncorruptible God into an image made like to corruptible man, and to birds, and

fourfooted beasts, and creeping things. [before there was philosophical naturalism idolatry rather than naturalism followed intellectualism]

Wherefore God also gave them up to uncleanness through the lusts of their own hearts, to dishonour their own bodies between themselves: [the corruption of mankind]

Who changed the truth of God into a lie, and worshipped and served the creature more than the Creator, who is blessed for ever. Amen.

For this cause God gave them up unto vile affections: for even their women did change the natural use into that which is against nature:

And likewise also the men, leaving the natural use of the woman, burned in their lust one toward another; men with men working that which is unseemly, and receiving in themselves that recompence of their error which was meet.

And even as they did not like to retain God in [their] knowledge, God gave them over to a reprobate mind, to do those things which are not convenient;

Being filled with all unrighteousness, fornication, wickedness, covetousness, maliciousness; full of envy, murder, debate, deceit, malignity; whisperers,

Backbiters, haters of God, despiteful, proud, boasters, inventors of evil things, disobedient to parents,

Without understanding, covenant breakers, without natural affection, implacable, unmerciful:

Who knowing the judgment of God, that they which commit such things are worthy of death, not only do the same, but have pleasure in them that do them.

This Scripture is about the pre-Abraham days when, the pre-history ancient Greeks (and all peoples of the earth such as the Americans, Europeans, Chinese, Koreans, Japanese, Indians, Africans, Middle East people, etc, all of whom once knew God), turned away from God. Because our ancestors did not want to retain God in their knowledge they found a world view that allowed them to understand that idol gods were the real gods.

Isaiah 44:9 says of the man who carves an idol from part of a tree: "And none considereth in his heart, neither [is there] knowledge nor understanding to say, I have burned part of it in the fire; yea, also I have baked bread upon the coals thereof; I have roasted flesh, and eaten [it]: and shall I make the residue thereof an abomination? shall I fall down to the stock of a tree?" Scripture is saying that it does not matter how many people practice idolatry, we should not try too hard to understand their world view because all people should know better. Perhaps we can allow a little cultural relativism. Maybe God was a bit softer on idolatry for gentiles who were not in contact with Israel because Acts 17:29-30 says: "For as much then as we are the offspring of God, we ought not to think that the Godhead is like unto gold, or silver, or stone, graven by art and man's device. And the times of this ignorance God winked at; but now commandeth all men every where to repent:"

In our modern world we have philosophical naturalism instead of idols. Philosophical naturalism causes science to suppress the idea that there is a real creator. To say that philosophical naturalism is spiritual error is to say that if God suddenly came out of hiding, we would see something insane about our old world view just as idol worshippers would understand that there is something insane about worshipping an idol. Under other circumstances though, philosophical naturalism as an external influence will be invisible to us because we are in agreement with it.

2 Thessalonians 3:2 says: "And that we may be delivered from unreasonable and wicked men: for all [men] have not faith." According to this Scripture we should not try too hard to understand the error of the God-is-dead enlightenment because an absence of faith is an unreasonable and wicked error. Hebrews 3:12 similarly warns Christians: "Take heed, brethren, lest there be in any of you an evil heart of unbelief, in departing from the living God." Philosophical naturalism is a western version of unbelief.

According to the above Romans Scripture all who do not believe in a creator are without excuse. In fact cultures in contact with the Bible cannot even be excused for believing that mice can grow from wheat grains even though this seems to be a minor addition to the absurd idea that wheat can grow from wheat grains. See Genesis 1:11-12 and 1 Corinthians 15:37. However we do not need the Bible to understand that their was no excuse for believing in spontaneous generation and spontaneous generation was based on the same kind 'evil heart of unbelief' as philosophical naturalism.

Both Spinoza and Holbach used Hobbes' concept of familiarity to convey the idea that naturalism is all there is. Although familiarity for things like wheat growing from a wheat grain should be a source of inspiration for a belief in God it has the reverse effect on those who 'do not wish to retain God in their thoughts' as per the above Romans Scripture. This is because when we do not retain God in our thinking, familiarity with things that grow teaches us that the slow changes (of evolution and spontaneous generation) are natural. If we look at a drawing of say the workings of an ear and we bring to mind that ear came from a single cell, (or if we think about wheat growing from a wheat grain) then we should be wondering how the creator of the universe could do that. We should not be telling ourselves that a trial-and-error algorithm did that.

Familiarity with the natural magic of wheat growing from a wheat grain 'without retaining God in their thoughts' (as per verse 28 in the above Romans Scripture) is why people were happy to believe in spontaneous generation and it is why Darwin was happy to include natural magic in his solution. In his *Why I Am Not A Christian*, Russell says regarding the design argument: "It is not that their environment was made to be suitable to them but that they *grew* to be suitable to it, and that is the basis of adaptation. There is no evidence of design about it." Russell sees no need to consider whether natural selection is up to the task because he became familiar with the fact that things can grow without retaining God in his thoughts.

As mentioned previously, there is also a cultural aspect to the error of philosophical naturalism. According to the above Romans Scripture, when everything was a mystery 'intellectualism' helped our ancestors to believe in idols rather than a creator. This interpretation comes from the words 'professing themselves to be wise'. Intellectualism lends its support to philosophical naturalism and it creates cultural pressures.

Intellectualism lives in the valley of naturalism adjacent the improbable peak of creationism. Some of us attempt to climb the peak of creationism but given the slightest challenge to creationism, such as if we find a reason to believe that one link in an unbelievable chain of ideas (say if we determine that the thirty or so elements necessary for life to exist are on earth — I am kidding; or if we can show that the twenty amino acids can be generated naturally — I am not kidding), then we will be nudged from the peak into the valley where we will chide ourselves for believing the superstitious nonsense of celestial spheres, ghosts, goblins and God. On page 155 of his *God Delusion*, Dawkins argues: "Those people who leap from personal bafflement at a natural phenomenon straight to a hasty invocation of the supernatural are no better than the fools who see a conjuror bending a spoon and leap to the conclusion that it is 'paranormal'." Philosophical naturalism is the gravity. Intellectualism explains the cultural pressures that Dawkins invokes.

Dawkins suggests that scientist who stand up for creationism will be made to feel intellectually ashamed. As it turns out though, the primary prejudice of science is a mistake. Spokesmen for evolution who leap from personal bafflement at natural creation straight to a hasty assumption that there must be a natural solution no matter how unlikely, are no better than scientists who leapt to the conclusion of spontaneous generation. In our current culture my argument could never make individual scientists feel ashamed. Instead intellectualism ensures that scientists are ashamed of believing in the Bible. Psalms 10:4 "The wicked, through the pride of his countenance, will not seek *after God* : God is not in all his thoughts." The intellectual countenance is sensitive to the fact that God can be put into the same class as ghosts and goblins -- apparently because all these words start with the letter 'g'.

We should spend more time contemplating creation — more time watching David Attenborough documentaries and less time listening to David Attenborough documentaries because God is not fond of intellectualism. Jesus said (Mark 8:38): "Whosoever therefore shall be ashamed of me and of my words in this adulterous and sinful generation; of him also shall the Son of man be ashamed, when he cometh in the glory ofhis Father with the holy angels."

And in Luke 10:21 Jesus said: "In that hour Jesus rejoiced in spirit, and said, I thank thee, O Father, Lord of heaven and earth, that thou hast hid these things from the wise and prudent, and hast revealed them unto babes: even so, Father; for so it seemed good in thy sight." The idea that there is a real creator has been hid from the wise and prudent for the last 150 years. Dawkins is helping Jesus with the polarisation of the wise and prudent by adding to the pressure of intellectualism.

Spokesmen for evolution also add to the importance of evolution. Spokesmen for evolution say that 'nothing in biology makes sense without evolution', 'America is in danger of loosing it's scientific edge', 'evolution is one of the half dozen "good ideas" developed by science', 'fundamentalist religion is hell-bent on ruining the scientific education of countless of thousands of innocent, well-meaning, eager young minds'.

According to Creation Ministries International, the BBC claims that 'intelligent design is at war with science'. Creation Ministries International quotes Dawkins as saying: "It is enormously damaging if the world's leading economic power were subverted and undermined by this kind of anti-scientific tendency" and Ken Miller as saying "I truly regard America's scientific soul to be at stake in this struggle, and losing that, I think would be a great tragedy." Apparently Dawkins and Miller are fundamentalists because here they are trying to incite the 'concerned spleen'. Popper's advice would have worked well against these arguments for concern.

As will be discussed in another book, Dawkins raises some concerns against the Bible and Christianity regarding the doctrine of hell. Dawkins says that the doctrine of hell can cause torment in the mind of people who were once Christians. And even though such concerns are very similar to Epicurus' original concerns, unlike the 'importance of science concerns' they are real concerns. However, Dawkins perspective on hell is not the only logical perspective.

If God did create then God is a real God who can do things. If there is a real God then maybe there is a real hell and a real heaven. Given the doctrine of hell and a little doubt, science should be able to see the need for some conservative bias and responsibility in what we teach our school children. Responsibility means that unless your scientific opinion about the existence of a real God is like my opinion about the Book of Mormon and the Koran, then your arguments should not be made as persuasive as possible, because they might be wrong. If your son or daughter was on their death bed you might not be so eager to teach them that Jesus was not who he claimed to be, even if you did not believe in Jesus. Therefore, if we are going to teach our children that there are no reasons for believing in a creator who can do things then we need to consider our arguments carefully. Because we find facts that can speak for themselves, we may not be able to pass the responsibility off onto the facts.

I take full responsibility for the judgement that the Book of Mormon and the Koran have nothing to do with God, but with the exception of a few people like Dawkins and Dennett, many spokesmen for evolution dodge their responsibility with fuzzy religious thinking. One dodgy argument says 'evolution is not anti-God or anti-religion, but that it is only anti-fundamentalism because so many religious people believe in god and evolution'. Of course there must be some people who claim to have some sort of belief in God and believe in evolution because evolution and God are not mutually exclusive beliefs like Christ and anti-Christ beliefs. Fundamentalism can be bad because the book of Job tells us that God had a problem with people who were defending their religion. However, fundamentalism can also be good because God is a logical God. For example, 1 Corinthians 15:14 -19, tells us that God has more respect for Dawkins thinking than for fuzzy religious thinking:

> And if Christ be not risen, then [is] our preaching vain, and your faith [is] also vain. Yea, and we are found false witnesses of God; because we have testified of God that he raised up Christ: whom he raised not up, if so be that the dead rise not. For if the dead rise not, then is not Christ raised: And if Christ be not raised, your faith [is] vain; ye are yet in your sins. Then they also which are fallen asleep in Christ are perished. If in this life only we have hope in Christ, we areof all men most miserable.

The Bible is absolutely dogmatic about who Jesus was. The Bible gives us no choice but to conclude that either Jesus was the truth or he was not, with no hope for any in between fuzzy truth. According to the Gospels: "Jesus saith unto him, I am the way, the truth, and the life: no man cometh unto the Father, but by me." That is one of many extremely clear dogmatic assertions about who Jesus is. Either they are true or they are not.

If Muhammad was a prophet and the Koran is the word of God then we can be sure that Jesus did not rise from the dead. If Jesus did not truly rise from the dead then what is the point of Christianity?

Fuzzy irresponsible thinking says that science is not anti-God but it is only anti-fundamentalism. Clear responsible thinking says that evolution is against religion as taught and practiced by Jesus and the apostles and modern day fundamentalists, but it is not against religion made by man under the inspiration of evolution. Clear responsible thinking understands that evolution asserts to our children that 'there is no evidence for a real intervening God' because that is exactly what it does.

Clear responsible thinking requires that we remind ourselves over and over and over again that if we remove the bias of philosophical naturalism, then what we really have is an invalid faith in philosophical naturalism (i.e. a delusion) versus the difficulty of the problem of creation. We should not be assuming naturalism even if creation was not such a difficult problem. Because it is such a difficult problem, we should all be able to say that evolution has absolutely no chance of being the truth and any claim that the fossil record tells us that evolution happened should be seen as an empirical contradiction because we should need an incredible amount of empirical evidence to even consider that naturalism can make walking, talking, thinking, learning, self-maintaining, self-reproducing biochemical machines. We should be supposing that the evolution algorithm is incapable of the creation of complexity, and we should be supposing that the problem of creation of the first life form is just like it seems except that our evil heart of unbelief tells us that that all problems can be solved by naturalism.

Except that we came to understand the world 'without retaining God in our thoughts' we should all be confronted by an irresistible argument for a real intervening creator rather than the remote, semi-real observer preached by science and religion. Instead of intellectual science being the inspiration for 'I do not believe in God because I do not believe in Mother Goose', intelligent science should be the inspiration for 'I believe in God because I do not believe in the fairy-god-mother'.

I have read the end of The Book, and the end of The Book tells me that sooner or later God is going to come out of hiding and science will have no choice but to change. However, we do not need to wait until God comes out of hiding to insist that science has an ordinary unscientific understanding of truth. We can insist that science preaches what the arguments seem to say about the theory of evolution. And regarding scientific bodies funded by the government, we can enforce an ordinary understanding of the truth. Creationists can help this process along by politely asking appropriate questions. If we do that then philosophical naturalism may loosen its hold on Australia. This may help to revive pre-Darwin God-thinking.

An enlightenment of common sense will make space for an enlightenment of the supernatural. According to 2 Kings 6:14-17: "Therefore sent he thither horses, and chariots, and a great host: and they came by night, and compassed the city about. And when the servant of the man of God was risen early, and gone forth, behold, an host compassed the city both with horses and chariots. And his servant said unto him, Alas, my master! how shall we do? And he answered, Fear not: for they that be with us are more than they that be with them. And Elisha prayed, and said, LORD, I pray thee, open his eyes, that he may see. And the LORD opened the eyes of the young man; and he saw: and, behold, the mountain was full of horses and chariots of fire roundabout Elisha." We live in a spiritual world where our only weapons are things like faith, truth and righteousness. (Some readers may note that there is a gap in my explanation, namely that I have avoided the issue that there is no mention of dinosaurs in the bible. In very early drafts of this book what is now the gap in this discussion, was the core of what I wanted to say but I have reduced my explanation to a few paragraphs in the next book. There are many other questions that Dawkins raises against the Bible, which have yet to be considered.)

NOTES

1. GRANTS ACCOUNT OF A DISPUTE BETWEEN THE CATHOLIC CHURCH AND PHILOSOPHERS

The dispute is appalent in at least three major controversies, which involved (1) the eternity of the world, (2) the so-called doctrine of the double truth, and (3) God's absolute power. Interdisciplinary friction between theologians and natural philosophers was compounded by intradisciplinary rivalries among the theologians themselves. Neoconservative Augustiniaris were pitted against the Dominican followers of Saint Thomas Aquinas, with the former concerned about the overreliance of the Dominicans on Aristotelian philosophy and the latter determined to pursue a harmonization of reason and revelation. The condemned articles themselves, however, serve as good illustrations of the controversies that loomed large in the late thirteenth century.

The following three articles convey the flavor of the hostility between arts masters and theologians:

152. That theological discussions are based on fables.

153. That nothing is known better because of knowing theology.

154. That the only wise men of the world are philosophers....

Of the three major issues described earlier, the third, the challenge to God's absolute power, may have been regarded as the most potentially subversive of theological traditions. Scattered through the works of Aristotle were propositions and conclusions demonstrating the natural impossibility of certain phenomena. For example,

Aristotle had shown that it was impossible for a vacuum to occur naturally inside or outside the world and had also demonstrated the impossibility that other worlds might exist naturally beyond ours. Theologians came to view these Aristotelian claims of natural impossibility as restrictions on God's absolute power to do as he pleased. Why should God not be able to produce a vacuum inside or outside the world if he chose to do so? Why could he not create other worlds if he chose to do so? Article 147 reveals the attitude of the bishop of Paris and his colleagues when it denounced as erroneous the opinion that not even God could do what was naturally impossible. The following articles in the Condemnation of 1277 were among those thought to place limitations on God's absolute power to do whatever he pleases:

21. That nothing happens by chance, but all things occur from necessity and that all future things that will be will be of necessity, and those that will not be it is impossible for them to be ....

22. That the first cause [that is, God] could not make several worlds.

23. That without a proper agent, as a father and a man, a man could not be made by God[alone].

24. That God cannot be the cause of a new act [or thing], nor can He produce something anew.

25. That God could not move the heavens [or world] with a rectilinear motion; and the reason is that a vacuum would remain.

26. That an accident existing without a subject is not an accident, except equivocally; [and] that it is impossible that a quantity or dimension exist by itself because that would make it a substance.

27. That to make an accident exist without a subject is an impossible argument that implies a contradiction.
28. That God cannot make an accident exist without a subject, nor make several dimensions exist simultaneously [in the same place].

Many more articles limiting God's power could be cited. All were condemned because the theological authorities wanted everyone to concede that God could do anything whatever short of a logical contradiction. By condemning the opinion that God could not create other worlds, article 34 made it mandatory to concede that he could create as many as he pleased. Although no one was required to believe that God had created other worlds, and no one is known to have so believed, the effect of article 34 on natural philosophy was to encourage speculation about the conditions and circumstances that would obtain if God had indeed created other worlds. Article 49 denied to God the capability of moving the outermost heaven, and therefore the world itself, with a rectilinear motion, because such a motion would leave behind a vacuum after the departure of the world from its present position. Following the condemnation of article 49 in 1277, scholastic natural philosophers routinely conceded that if it pleased God to do so, he could indeed move the world rectilinearly.

In articles 139, 140, and 141, the authorities condemned the seemingly self-evident Aristotelian principle that an accident could not exist without a subject, or substance, in which to inhere and that God could not create an accident, or quality, that did not inhere in a subject or substance. They further condemned (in articles 139 and 141) the Aristotelian axiom that no quantity or dimension could exist independently of a material body and also denounced the equally fundamental Aristotelian principle that two or more dimensions could not exist simultaneously in the same place. Not only did articles 139, 140, and 141 qualify as placing limitations on God's power, but they denied to God the power to effect the theological dogma of the Eucharist, whereby God miraculously produces the actual body and blood of Christ in the bread and wine used in the mass. Although the bread has become the substance of the body of Christ, the bread itself remains with its usual visible, accidental properties. Those accidents, however, no longer inhere in the substance of the bread. And because their existence in Christ is denied, one must conclude that the accidents of the bread do not inhere in any substance, a conclusion that is contrary to Aristotle's doctrine that all accidents must inhere in a substance. In the fourteenth century, Walter Burley argued that just as (in the Eucharist) God makes a quantity devoid of corporeal substance, so also could he make a quantity devoid of inhering qualities, by which Burley meant an extended vacuum through which light and heavy bodies could move. Thus did Burley link the condemned articles on the supernatural separation of accidents from their subjects, or attributes from their substances, with a much-discussed medieval problem of motion in a separate, empty space.

Article 48 was also condemned because it placed restrictions on God's absolute power.The concept of a lawful, regular, and even deterministic world held appeal for astrologers and was also basic to Aristotle's natural philosophy. The idea that God would intervene miraculously in the natural order and do something new was alien to Aristotle's physics but was an important feature of the Christian religion. Medieval followers of Aristotle's philosophy were now warned that they could not deny the occurrence of such actions.

2. ARISTOTLE COMMENTS ON SCIENTIFIC KNOWLEDGE:

The following is a selection of some things that Aristotle's *Posterior Analytics* had to say about how we learn about the world we live in:

> 1.1. All instruction given or received by way of argument proceeds from pre-existent knowledge. This becomes evident upon a survey of all the species of such instruction. The mathematical sciences and all other speculative disciplines are acquired in this way, and so are the two forms of dialectical reasoning, syllogistic and inductive; for each of these latter make use of old knowledge to impart new, the syllogism [deductive, demonstrative or tautological reasoning] assuming an audience that accepts its premises [the givens], induction exhibiting the universal as implicit in the clearly known particular. Again, the persuasion exerted by rhetorical arguments is in principle the same, since they use either example, a kind of induction, or enthymeme, a form of syllogism.
>
> 1.3. Some hold that, owing to the necessity of knowing the primary premises, there is no scientific knowledge. Others think there is, but that all truths are demonstrable. Neither doctrine is either true or a necessary deduction from the premises. The first school, assuming that there is no way of knowing other than by demonstration, maintain that an infinite regress is involved, on the ground that if behind the prior stands no primary, we could not know the posterior through the prior (wherein they are right, for one cannot traverse an infinite series): if on the other hand — they say — the series terminates and there are primary premises, yet these are unknowable because they are incapable of demonstration, which according to them is the only form of knowledge. And since thus one cannot know the primary premises, knowledge of the

conclusions which follow from them is not pure scientific knowledge nor properly knowing at all, but rests on the mere supposition that the premises are true. The other party agree with them as regards knowing, holding that it is only possible by demonstration, but they see no difficulty in holding that all truths are demonstrated, on the ground that demonstration may be circular and reciprocal.

Our own doctrine is that not all knowledge is demonstrative: on the contrary, knowledge of the immediate premises is independent of demonstration. (The necessity of this is obvious; for since we must know the prior premises from which the demonstration is drawn, and since the regress must end in immediate truths, those truths must be indemonstrable.) Such, then, is our doctrine, and in addition we maintain that besides scientific knowledge there is its originative source which enables us to recognize the definitions...

2.19. We have already said that scientific knowledge through demonstration is impossible unless a man knows the primary immediate premises...

So out of sense-perception [sense perception provides experience with the world] comesto be what we call memory, and out of frequently repeated memories of the same thing develops experience; for a number of memories constitute a single experience...

We conclude that these states of knowledge are neither innate in a determinate form, nor developed from other higher states of knowledge, but from sense-perception. It is like a rout in battle stopped by first one man making a stand and then another, until the original formation has been restored. The soul is so constituted as to be capable of this process...

> Thus it is clear that we must get to know the primary premises by induction; for the method by which even sense-perception implants the universal is inductive.
>
> 2.7 How then by definition shall we prove substance or essential nature? We cannot show it as a fresh fact necessarily following from the assumption of premises admitted to be facts—the method of demonstration: we may not proceed as by induction to establish a universal on the evidence of groups of particulars which offer no exception, because induction proves not what the essential nature of a thing is but that it has or has not some attribute. Therefore, since presumably one cannot prove essential nature by an appeal to sense perception or by pointing with the finger, what other method remains?

I am not sure what Aristotle means by his claim that 'there is an originative source which enables us to recognize the definitions' in the last line of 1.3 above, but most other things seem to make some sense. What Aristotle has to say seems more impressive than ideas like 'rest is the opposite of movement, and nothing moves by virtue of its own nature towards its opposite or own destruction' that came 2000 years later. On pages 68 and 69 of *The Foundations of Modern Science in the Middle Ages* Grant writes of Aristotle's influence:

> In much, if not most, of what he said, however, Aristotle's ideas were utilized as the best and most reliable guides to the comprehension of nature and its works. To medieval scholars, he was truly the Philosopher. In his commentary on Aristotle's On the Heavens (De caelo), Averroes paid Aristotle the highest possible tribute, declaring that Aristotle was a rule and exemplar which nature devised to show the final perfection of man ... the teaching of Aristotle is the supreme truth, because his mind was the final expression of the human mind. Wherefore it has

been well said that he was created and given to us by divine providence that we might know all that is to be known. Let us praise God, who set this man apart from all others in perfection, and made him approach very near to the highest dignity humanity can attain."

David Knowles, a historian of medieval philosophy, was not exaggerating when he called this "the most impressive eulogium ever given by one great philosopher to another." Indeed, Averroes considered Aristotle to be virtually infallible because in over one thousand years no error had been detected in his writings.

Aristotle was also greatly admired in the Latin West. Dante spoke for many when he described Aristotle as "the Master of them that know." Thomas Aquinas regarded Aristotle as someone who had attained the highest possible level of human thought without benefit of the Christian faith.

3. MATHEMATICAL CERTAINTY:

It may be that some of the difficulties in understanding what Aristotle has to say about scientific knowledge are due to a failure of the ancients to understand that we can be practically certain of some things (say the law that rocks and other heavy things cannot float above the earth), and yet, still agree that our knowledge is only probable knowledge. Aristotle may have thought that knowledge is only true knowledge if it is certain, but in everyday language, certainty does not always mean mathematical certainty.

In theory it should not be possible to say that creationism is not science except by use of some definition. However, in everyday communication, a 'not' is not always a 'logical not'. In order to make sense of the 'creationism is not science' assertion without the help of a definition, we must interpret 'not' as 'probably not'. And using 'not' for 'probably not' can be acceptable for the purposes of normal communication — we might say that 'the theory that there are fairies in the garden is not science' even without the help of a definition. Although almost all westerners use philosophical naturalism to say that creationism is not science in the sense that creationism is sufficiently unlikely, it is also the case that scientists want to say something derogatory and categorical about creationism, and to do that they also need to have a definition in mind.

One type of 'certainty' and one type of 'not science' is extinguished by even the most preposterous doubt, whereas the other is not. Probability values will not be as misleading if instead of using probability numbers from 0 to 1 we used a probability number Pn from + Infinity to – Infinity by converting probability to odds and taking the log (to base 2 say) of the number.

4. POPPER REJECTS ANY POSSIBILITY OF INDUCTIVE INFERENCES:

Popper liked to say things like 'numerous confirming instances of white swans do not tell us that all swans are white'. Except for those laws of science that can be justified with a definition, we know that there can never be such a thing as a law of science that always remains true. This is because God can do whatever he likes in his universe. This means that there is no way of getting from the observation that 'all swans have been white' to the conclusion that 'all swans are white' and therefore the problem of induction comes down to whether or not we can use induction to make a probable inference about a finite number of instances. In the following argument, taken from page 103 of *Popper Selections*, edited by David Miller, Popper argues that there is no such thing as inductive thinking even when we only have in mind a limited number of unobserved instances:

By an inductive inference is here meant an inference from repeatedly observed instances to some as yet unobserved instances...

There are various other philosophers who also regard as misconceived this traditional problem of induction. Some say that it is misconceived because no justification is needed for inductive inference; no more in fact than for deductive inference. Inductive inference is inductively valid just as deductive inference is deductively valid. I think it was Professor Strawson who was the first to say this.

I am of a different opinion. I hold with Hume that there simply is no such logical entity as an inductive inference; or, that all so-called inductive inferences are logically invalid — and even inductively invalid, to put it more sharply. We have many examples of deductively valid inferences, and even some partial criteria of deductive validity; but no example of an inductively valid inference exists. And I hold, incidentally, that this result can be found in Hume, even though Hume, at the same time, and in sharp contrast to myself, believed in psychological power of induction; not as a valid procedure, but as a procedure which animals and men successfully make use of, as a matter of fact and of biological necessity.

I take it as an important task to make clear, even at the cost of some repetition, where I agree and where I disagree with Hume.

I agree with Hume's opinion that induction is invalid and is in no sense justified. Consequently neither Hume nor I can accept the traditional formulations which uncritically ask for the justification of induction; such a request is uncritical because it is blind to the possibility that induction is invalid in every sense, and therefore unjustifiable.

> I disagree with Hume's opinion (the opinion incidentally of almost all philosophers) that induction is a fact and in any case needed. I hold that neither animals nor men use any procedure like induction, or any argument based on the repetition of instances....
>
> Thus it is not because induction can manage without justification that I am opposed to the traditional problem; on the contrary it would urgently need justification. But the need cannot be satisfied. Induction simply does not exist and the opposite view is a straightforward mistake.

5. EXACT DATA & HERMAN WEYL'S DELIBERATIONS:

As discussed in Section 2.3, Popper and Herman Weyl note that both interpolation and extrapolation of data is an attempt to make predictions about points for which we have no data. Although we can draw any number of curves between say twenty points that lie on a straight line, something similar to the linear curve (within the limits of error) would be more probable than some specific curve that oscillates like a sinusoidal curve but happens to intersect the points that lie on a straight line. Nevertheless, even if we insist on no special points (no piecewise smooth functions) we cannot eliminate the possibility of deviations (say in the form of bell curves) superimposed on the straight line. This is because we can confine the non-conformities to a sufficiently narrow portion of the domain such that we can say it is then unlikely that we would have picked up such behaviour in our sample of points. We know from past experience that we do not expect to encounter such deviations and we also know that every specific alternative to the straight line conjecture is nothing more than a wild guess.

Information theory shows us how to distinguish a coefficient from a constant. We could say that a constant was any number that could be expressed as a finite number of digits or a finite algorithm. In other words, a constant could be a number that can be expressed by a finite amount of information and a coefficient is any number that was not constant. This means that numbers like $1/3 = 0.3333...$ the $\sqrt{2} = 1.414...$, $\pi = 3.14159...$ are constants because they can be compressed into a finite amount of information. Now suppose, we select a random X = A and get Y = A from a unknown law. We will then have one datum point that coincides exactly with the point [A,A] where 'A' is an infinite sequence of digits which are not compressible into an algorithm (i.e. a coefficient point). When we have approximate data, we would normally say we need three points to even suggest a straight line (unless our recent experience with data suggests otherwise), but when the data is exact, because the accidental generation of the point [A,A] is impossible except if we have a law, then in fact only one point tells us that we have a straight line, which points towards the origin and has a slope of 1:1. With reference to Weyl's criteria to have a class of preselected functions, we can now have a countably infinite (or denumerably infinite) number of different functions in our class of pre-selected simple functions, and some 10 digit constant would apparently be no more ad hoc than the 1 in x^1 because we have a method of distinguishing coefficients from constants. Nevertheless, we could never check an infinite series of digits and therefore, we would have to rely on induction to conclude that two numbers were identical.

6. LARGE NUMBERS :

The haemoglobin number is generated from the number of ways of arranging the 20 amino acids to form the specific beta haemoglobin sequence of 146 amino acids and is therefore given by 20^{146}, which is approximately equal to $8.92*10^{189}$ and which is close enough to 10^{190}. An interesting feature of such large numbers is that when we multiply them by large numbers, they hardly seem to get any bigger. Multiplying the haemoglobin number by 10 billion hardly seems to make any difference. Of course, multiplying by 10 billion does make a difference but it is a fact that the enormous difference between 10^{160} and 10^{170} is not always obvious, except that when we subtract two very large numbers that seem nearly equal, the result will equal the larger number.

Just how large is the haemoglobin number? Loschmidt's number, which is about $2.686*10^{25}$, gives us the number of molecules in one cubic metre of perfect gas at standard conditions — we know that this represents a lot of molecules because they are able to maintain a perfectly uniform pressure on the sides a vessel through the average effect of the impact of the atoms with the sides of the vessel. So, from that we can appreciate that there are a lot of atoms in the earth and moreover, that the sun and stars are typically hundreds of thousands times larger than the earth and there are a lot of stars in the observable universe, and so we might be surprised to learn that the number of electrons in the observable universe is only about 10^{80}. Even if we imagined a observable universe for every electron in our observable universe and we counted the number of electrons in these 10^{80} observable universes, we would not reach the haemoglobin number. By some standards the haemoglobin number at 10^{190} appears to be a large number. However, numbers like the haemoglobin number are easy to generate with seemingly 'moderate' probability problems.

Cairns Smith likens the chances of making a batch of primed nucleotides by accident as similar to getting 140 sixes in a row from a throw of a dice. He then says that if we were to throw our die every second for say the 3.5 billion years of earth's history, then we would need far more die than there are electrons in the observable universe in order to be able to expect to generate the 140 sixes.

7. TABLE OF PROBABILITY VALUES:

The following table provides a summary of some probability values for the scientist, evolutionist and the biblical creationist. The objective of this table is to explore how background probabilities effect our thinking. The vertical line '|' means 'given', 'E' means 'evolution' in the sense of slow progressive changes (by whatever process), 'N' means 'naturalism-only' which is philosophical naturalism, 'C' means a 'creator' and the braces '{ }' mean 'not' whatever is in the braces. Where a value of 1 or 0 is indicated, this is not necessarily exactly 1 or 0, because the huge differences in probabilities, which would be visible in the 'log of the odds scale', have been concealed.

Some of the probability values are more or less by definition. The table assumes that design means God. For example, the probability of naturalism-only given not-evolution is defined as being zero i.e. $P(N|\{E\}) = 0$, and in this case we can say exactly zero since fast changes would be a miracle. Similarly if we have naturalism-only then the probability of evolution is defined as being exactly equal to 1 i.e. $P(E|N) = 1$.

TABLE OF PROBABILITY VALUES

PROBABILITY	NATURALIST	SCIENTIST	BIBLICAL CREATIONIST
P(E)	1 (Calculated)	0 (Fundamental)	0 (Fundamental)
P(N)	1 (Fundamental)	0 (Calculated)	0 (Fundamental)
$P(E\mid N)$	1 (Defined)	1 (Defined)	1 (Defined)
$P(N\mid E)$	1 (Defined)	1 (Defined)	0 - 1 ?
$P(E\mid \{N\})$	0 (Fundamental - probability of evolution withbackground information removed)	0 (Fundamental - probability of evolution withbackground information removed)	0 (Fundamental - probability of evolution withbackground information removed)
$P(N\mid \{E\})$	0 (Defined)	0 (Defined)	0 (Defined)
$P(C\mid E)$	0	0	0 - 1 ?
$P(C\mid \{E\})$	0 - 1 ?	1	1

Note that the non-biblical creationist will have the same probability values as the scientist and that even the biblical-creationist will have many of the same probability values as the scientist. This is because the scientist and the creationist should both agree that the probability of evolution, if we ignore the fact that life exists, is zero. In this case we are all making the judgment with the same background information which means we have removed the reasons for the judgements to be different. Providing there are no motivating factors and no errors, the judgments should be the same. Except that we need evolution to solve the problem then, evolution is unlikely or $P(E|\{N\}) = 0$.

The scientist calculates the probability of naturalism based on his fundamental assessment of evolution. Suppose the scientist thinks that there is no natural explanation for creation, the scientist will then have a reason to believe in creationism. The scientist may be willing to concede naturalism-only if he considers that naturalism will solve the creation problem, but he will also be willing to consider creationism if he sees that naturalism is not up to the task. The scientist and the creationist (but not the biblical creationist and the anti-creationist) are the true freethinkers.

Where the table indicates a range of values from 0 to 1, this indicates some confusion. For example if creationism is falsified (and therefore evolution is true), our typical biblical creationist would be confused because he may not know whether he should discard his fundamental probabilities. The decision would depend on the individual and how he judges the test results. If evolution was falsified, then the naturalist would be confused.

Although it is not obvious how much of our thinking comes from background probabilities we can create scenarios that remove the 'given' from our thinking. For example, suppose that we are some kind of observers living in God's presence and God conjectures to us the theory of evolution and asks us to place bets on whether life can be produced by such a theory. In this scenario, we should not be thinking from the perspective that 'it must have happened by naturalism therefore it happened by naturalism'. God says to us 'there are a few billion planets over there and I am going to let a billion years go past and I have given you everything for the process to occur — how do you bet on the chances of a biological machine assembling itself by accident and then going on from there to improve itself by mutations and natural selection?' We would all say that there is a zero probability of that happening because in this scenario, the naturalist is not a naturalist. If readers find it difficult to agree with this judgement than in my view, that is because more than a little effort is required to remove the understanding that 'evolution happened' from our mental state. Exposing the error of philosophical naturalism should also achieve this mental state, except that philosophical naturalism resists the understanding that there are no reasons to trust in philosophical naturalism.

Suppose that God conjectures to us Darwin's theory of evolution. This time, we must consider that Darwin believed that we could have a natural process of evolution without failure — that is without mutations going wrong. Darwin knew that there were failed mutations (birth defects), but as far as his process was concerned he only believed in good, better and best because he believed that the only differences between evolution from a fish to man and artificial selection of a better breed of life stock were the time scale and the objectives of the selector. So for Darwin, we must conjecture a theory of evolution that effectively worked without trial and error, given the evidence of artificial selection. Darwinism uses trial-and-error to select the best selection but not to create the best selection. Without a faith in a some kind of natural magic Darwin's idea does not work. Darwin's idea only seems to work as a type of creationism that maintains that all life is designed with the ability to evolve like a seed is designed with the ability to grow. Darwin's evolution was similar to spontaneous generation and ancient Greek atomism.

Or suppose that in coming days, people of earth will get to see an angel flying around the earth preaching the gospel and another angel flying over the anti-Christs kingdom warning people not to take the mark of the beast. And suppose that all kinds of supernatural demonic monsters were released to afflict mankind. In such circumstances, people like Holbach will continue to believe in the laws of physics and chemistry, but they will doubt philosophical naturalism. Or let's suppose that Darwin went to hell immediately when he died. Not only would he immediately discard his theory of evolution, but he would very quickly understand how this thinking in terms of naturalism led him astray because he would very quickly understand that apart from the fact that he chose not to retain God in his thinking, there is nothing in the world of science that suggests that he should have extended the natural laws to explain creation.

The naturalist says that the probability of naturalism-only is 1 (i.e. P(N)= 1) based on the fact that naturalism explains everything. The naturalist cannot produce an argument for naturalism-only and his thinking is impossible to justify, given that creation is a special problem. For our anti-creationist, the probability of evolution is not a fundamental probability, but it is calculated on the basis of his faith in naturalism and the fact that there only seems to be one possible natural explanation (Of course all such calculations are subjective large number calculations that take place in our central processor in a fraction of a second).

The naturalist is fundamentally stuck on naturalism and he is stuck on evolution in the sense of slow progressive changes.

He will also be stuck the mutation-selection-copy process because he will understand that their cannot be a creation law of the universe. Suppose though that science discovered an alternative process that was clearly more likely to satisfy the creation problem — the alternative theory would allow naturalists to look at the mutation-selection-copy process in a critical perspective, rather than a perspective that the mutation-selection-copy process must be right, because it is the only natural possibility.

Imagine yourself in God's presence as some kind of being, conversing with God. God says: "My plan is to create life that can reproduce - what do you think of my plan". You say: "Do you mean that you are you going to give life your power to create? That is an amazing plan." God says: "Not really. They will have the power to create but they will not have my power to create." Readers should be able to finish off this conversation with God for themselves and in the process they should understand that we life in a world of utter mystery. So we understand a few supernatural laws that tell us how the planets move. So we no longer imagine crystalline spheres above us supporting and moving the planets. Such things do not compare with the creation of biochemical machines that can reproduce.

Because we know that we are here, people who believe in God are compelled to believe that God can make the biochemical machines we see around us whereas the fact that God hides himself means that freethinkers are compelled to believe that a logical algorithm can make the biochemical machines we see around us.

8. RUSSELL'S CRITICISM OF GREEK SCIENCE :

As mentioned in section 5.1 Empedocles developed an early form of the theory of evolution. On pages 72 and 75 of his *History of Western Philosophy*, Russell argues that

Empedocles philosophy was 'more scientific' than that of Parmenides, Plato, and Aristotle:

> His most important contribution to science was his discovery of air as a separate substance. This he proved by the observation that when a bucket or any similar vessel is put upside down into water, the water does not enter into the bucket. He says: 'When a girl, playing with a water-clock of shining brass, puts the orifice of the pipe upon her comely hand, and dips the water clock into the yielding mass of silvery water, the stream does not then flow into the vessel, but the bulk of the air inside, pressing upon the close-packed perforations, keeps it out till she uncovers the compressed stream; but then air escapes and an equal volume of water runs in.'
>
> This passage occurs in an explanation of respiration.
>
> He also discovered at least one example of centrifugal force: that if a cup of water is whirled round at the end of a string, the water does not come out.
>
> He knew that there is sex in plants, and he had a theory (some what fantastic, it must be admitted) of evolution and the survival of the fittest. Originally, 'countless tribes of mortal creatures were scattered abroad endowed with all manner of forms, a wonder to behold.' There were heads without necks, arms without shoulders, eyes without foreheads, solitary limbs seeking for union. These things joined together as each might chance; there were shambling creatures with countless hands, creatures with faces and breasts looking in different directions, creatures with the bodies of oxen and the faces of men, and others with the faces of oxen and the bodies of men. There were hermaphrodites combining the natures of men and women, but sterile. In the end, only certain forms survived.

> As regards astronomy: he knew that the moon shines by reflected light, and thought that this is also true of the sun; he said that light takes time to travel, but so little time that we cannot observe it; he knew that solar eclipses are caused by the interposition of the moon, a fact which he seems to have learnt from Anaxagoras.
>
> He was the founder of the Italian school of medicine, and the medical school which sprang from him influenced both Plato and Aristotle. According to Burnet (p. 234), it affected the whole tendency of scientific and philosophical thinking.
>
> All this shows the scientific vigour of his time, which was not equalled in the later ages of Greece.
>
> I come now to his cosmology. It was he, as already mentioned, who established earth, air, fire, and water as the four elements (though the word 'element' was not used by him). Each of these was everlasting, but they could be mixed in different proportions and thus produce the changing complex substances that we find in the world. …
>
> The originality of Empedocles, outside science, consists in the doctrine of the four elements, and in the use of the two principles of Love and Strife to explain change.
>
> He rejected monism, and regarded the course of nature as regulated by chance and necessity rather than by purpose. In these respects his philosophy was more scientific than those of Parmenides, Plato, and Aristotle. In other respects, it is true, he acquiesced in current superstitions; but in this he was no worse than many more recent men of science.

Russell takes issue with the fact that the ancient Greeks used purpose (teleological thinking) to explain. Russell explains teleological thinking in his discussion of Leucippus and Democritus (~460~370 BC):

> [Leucippus and Democritus] The fundamental ideas of the common philosophy of Leucippus and Democritus were due to the former, but as regards the working out it is hardly possible to disentangle them, nor is it, for our purposes, important to make the attempt. Leucippus, if not Democritus, was led to atomism in the attempt to mediate between monism and pluralism, as represented by Parmenides and Empedocles respectively. Their point of view was remarkably like that of modem science, and avoided most of the faults to which Greek speculation was prone. They believed that everything is composed of atoms, which are physically, but not geometrically, indivisible; that between the atoms there is empty space; that atoms are in destructible; that they always have been, and always will be, in motion; that there are an infinite number of atoms, and even of kinds of atoms, the differences being as regards shape and size. Aristotle asserts that, according to the atomists, atoms also differ as regards heat, the spherical atoms, which compose fire, being the hottest; and as regards weight, he quotes Democritus as saying 'The more any indivisible exceeds, the heavier it is.' But the question whether atoms are originally possessed of weight in the theories of the atomists is a controversial one.
>
> The atoms were always in motion, but there is disagreement among commentators as to the character of the original motion. Some, especially Zeller, hold that the atoms were thought to be always falling, and that the heavier ones fell faster; they thus caught up the lighter ones, there were impacts, and the atoms were deflected like billiard balls. This was certainly the view of Epicurus, who in most respects based his theories on those of Democritus, while

trying, rather unintelligently, to take account of Aristotle's criticisms. But there is considerable reason to think that weight was not an original property of the atoms of Leucippus and Democritus. It seems more probable that, on their view, atoms were originally moving at random, as in the modem kinetic theory of gases. Democritus said there was neither up nor down in the infinite void, and compared the movement of atoms in the soul to that of motes in a sunbeam when there is no wind. This is a much more intelligent view than that of Epicurus, and I think we may assume it to have been that of Leucippus and Democritus

As a result of collisions, collections of atoms came to form vortices. The rest proceeded much as in Anaxagoras, but it was an advance to explain the vortices mechanically rather than as due to the action of mind.

It was common in antiquity to reproach the atomists with attributing everything to chance. They were, on the contrary, strict determinists, who believed that everything happens in accordance with natural laws. Democritus explicitly denied that anything can happen by chance.! Leucippus, though his existence is questioned, is known to have said one thing: 'Naught happens for nothing, but everything from a ground and of necessity.' It is true that he gave no reason why the world should originally have been as it was; this, perhaps, might have been attributed to chance. But when once the world existed, its further development was unalterably fixed by mechanical principles. Aristotle and others reproached him and Democritus for not accounting for the original motion of the atoms, but in this the atomists were more scientific than their critics. Causation must start from something, and wherever it starts no cause can be assigned for the initial datum. The world may be attributed to a Creator, but even then the Creator Himself is unaccounted for. The theory of the atomists, in fact, was more nearly that of modern science than any other theory propounded in antiquity.

The atomists, unlike Socrates, Plato, and Aristotle, sought to explain the world without introducing the notion of purpose or final cause. The 'final cause' of an occurrence is an event in the future for the sake of which the occurrence takes place. In human affairs, this conception is applicable. Why does the baker make bread? Because people will be hungry. Why are railways built? Because people will wish to travel. In such cases, things are explained by the purpose they serve. When we ask 'why?' concerning an event, we may mean either of two things. We may mean: 'What purpose did this event serve?' or we may mean: 'What earlier circumstances caused this event?' The answer to the former question is a teleological explanation, or an explanation by final causes; the answer to the latter question is a mechanistic explanation. I do not see how it could have been known in advance which of these two questions science ought to ask, or whether it ought to ask both. But experience has shown that the mechanistic question leads to scientific knowledge, while the teleological question does not. The atomists asked the mechanistic question, and gave a mechanistic answer. Their successors, until the Renaissance, were more interested in the teleological question, and thus led science up a blind alley... Democritus was a thorough-going materialist; for him, as we have seen, the soul was composed of atoms, and thought was a physical process. There was no purpose in the universe; there were only atoms governed by mechanical laws. He disbelieved in popular religion, and he argued against the nous of Anaxagoras. In ethics he considered cheerfulness the goal of life, and regarded moderation and culture as

> the best means to it. He disliked everything violent and passionate; he disapproved of sex, because, he said, it involved the overwhelming of consciousness by pleasure. He valued friendship, but thought ill of women, and did not desire children, because their education interferes with philosophy. In all this, he was very like Jeremy Bentham; he was equally so in his love of what the Greeks called democracy.

Russell's argument suggests that the ancient Greeks had a choice about asking the teleological question. We can't fail to notice the order in a quartz crystal and we can't fail to notice the order in the flat surface of water and we can't fail to notice the left hand looks like the right hand. Given a little understanding, we have no choice but to notice that a birds wings seem to be designed for flying, a fish gills seemed to be designed for breathing underwater, eyes seem to be designed for seeing, ears seem to be designed for hearing, hearts seem to be designed for pumping blood, etc. We can therefore say that observant Greeks, had no choice but to notice the purpose in design, because purpose is evident in design.

We need philosophical naturalism to ask the mechanistic question, but we have no choice but to ask the teleological question. In the case of a crystal and the flat water we can simplify the description using a geometrical pattern. In the case of life we can simplify the description using purpose. If we found a watch in some old rocks, then we would not seek a natural explanation for how the watch came to be. The purpose along with some laws of physics should describe the watch. The same argument can be applied to life except that Russell thinks that we should have sought a natural explanation.

It is true that Aristotle's celestial spheres do not seem very scientific. They seem to suggest that a creator put the spheres there just for the purpose of regulating the movement of the heavenly bodies, whereas Newton's laws tell us that the planets are kept in place by the same law that makes things fall down. Therefore there is a sense in which we can say that ancient Greek atomism, that required materials to have the propensity to organise themselves into complexity for a purpose, was more scientific than Aristotles teleological thinking.

Greek atomism may have been more scientific but it was not sensible. Materials might organise themselves into a perfect crystalline shape but we should not expect materials to organise themselves into a book or a machine regardless of how scientific the theory looks.

Although Aristotle's argument and intelligent design are generally identified as teleological arguments they are very different kinds of arguments. Regarding objects made by man such as knife or a kitchen table then of course Aristotle understood the need for an agent, because teleological thinking helped to explain how such things were brought into existence. However, in Aristotle's thinking, creation did not need an explanation because Aristotle assumed that creation always existed.

If we could embrace the idea that creation always existed then even people like Russell should be happy to categorise Aristotle's teleological thinking as scientific. Aristotle's teleological thinking seems unscientific because our current culture believes that the earth had a beginning. Aristotles crystalline sphere theory perhaps came from the fact that Aristotle was doing his thinking using the 'natural law' that everything could be explained using teleological thinking.

Because Aristotle was familiar with the concept of purpose in design (or purpose of the machine) he used purpose to explain things. Ancient science said things like 'nature abhors a vacuum', that 'rocks fall to the earth because they are like the earth' and that 'smoke rises because it is like the air'. All these laws seem to have an element of purpose in the sense that they are goal-seeking, and in theory, if that kind of thinking could be used to predict then that kind of thinking would be similar to Newton's laws. In the modern era we say that masses attract. More importantly, if we added the assumption that creation did not have a beginning, then the offence would disappear because under that scenario teleological thinking would not suggest that there was a real God who can do things.

We can also say that Aristotle's philosophy was scientific because like modern science, Aristotle had a god that didn't do anything. It seems that Aristotle, Empedocles and Anaxagoras had a religious form of philosophical naturalism, whereas Democritus (and Leucippus) had a strong form of philosophical naturalism.

If a man sold his house, car and all that he owned to buy lottery tickets to win a fortune, then we should not be commending the decision unless we can see something good behind the apparent insanity. This is to say that, even if it somehow it turned out that frogs can grow from mud, we should still say that success inadvertently came from foolishness.

However if a man who sold his possessions to buy lottery tickets was remarkably successful, say if he repeated this insanity many times and kept getting richer then we would be compelled to look harder to find what was good about his thinking. Russell sees the remarkable success of science and perhaps he understands the cause of that remarkable success to be the scientific attitude. If we think of the success of science in terms of 'fostering technological innovation, developing explanations for the changing of the seasons, the movements of the sun and stars, the structure of matter, the shaping of mountains and valleys, developing cures for diseases, and generating the knowledge needed to produce innumerable labour-saving devices' including all the advances in genetic science then the theory of evolution, even if it was good science does not belong in that category. With this measure of the success of science, we can say that if science had only applied Anaxagoras' naturalism for non-living things, then science would have been just as successful without being foolish.

9. ANCIENT ATTITUDES TO NATURALISM AND INTERVENTION:

Gazali (1058-1111), the fundamentalist Asharite Muslim who famously opposed the Islamic philosophers, supported the doctrine of occasionalism. Gazali's doctrine of occasionalism said that that there is no such thing as cause and effect because God is the cause of all things. The Asharites (named after Abu Hasan al-Ash'ari (873-935)) were a group of Muslims who asserted the Koran was not created but that it was always in existence with God and that there was no such thing as right and wrong except as God commands. On pages 166-170 of his *The Incoherence of the Philosophers*, Gazali writes:

> **[On causality and miracles]**
>
> (1) The connection between what is habitually believed to be a cause and what is habitually believed to be an effect is not necessary, according to us. But [with] any two things, where "this" is not "that" and "that" is not "this" and where neither the affirmation of the one entails the affirmation of the other nor the negation of the one entails negation of the other, it is not a necessity of the existence of the one that the other should exist, and it is not a necessity of the nonexistence of the one that the other should not exist—for example, the quenching of thirst and drinking, satiety and eating, burning and contact with fire, light and the appearance of the sun, death and decapitation, healing and the drinking of medicine, the purging of the bowels

and the using of a purgative, and so on to [include] all [that is] observable among connected things in medicine, astronomy, arts, and crafts. Their connection is due to the prior decree of God, who creates them side by side, not to its being necessary in itself, incapable of separation. On the contrary, it is within [divine] power to create satiety without eating, to create death without decapitation, to continue life after decapitation, and so on to all connected things. The philosophers denied the possibility of [this] and claimed it to be impossible.

(2) To examine these matters that are beyond enumeration will take a long time. Let us, then, take a specific example—namely, the burning of cotton, for instance, when in contact with fire. For we allow the possibility of the occurrence of the contact without the burning, and we allow as possible the occurrence of the Cotton's transformation into burnt ashes without contact with the fire. [The philosophers], however, deny the possibility of this.

(3) The discussion of this question involves three positions.

(4) The first position is for the opponent to claim that the agent the burning is the fire alone, it being an agent by nature [and] not by choice—hence, incapable of refraining from [acting according to] what is in its nature after contacting a substratum receptive of it. And this is one of the things we deny. On the contrary, we say:

(5) The one who enacts the burning by creating blackness in the cotton, [causing] separation in its parts, and making it cinder or ashes God, either through the mediation of His angels or without mediation. As for fire, which is inanimate, it has no action. For what proof is there it is the agent? They have no proof other than observing the occurrence of the burning at the [juncture of] contact

with the fire. Observation however, [only] shows the occurrence [of burning] at [the time of contact with the fire] but does not show the occurrence [of burning] [the fire] and [the fact] that there is no other cause for it. ... If, then, it is established that the Agent creates the burning through His will when the piece of cotton comes into contact with the fire, it becomes rationally possible [for God] not to create [the burning] with the existence of the contact.

(12) [To this] it may be said:

(13) This leads to the commission of repugnant contradictions. For if one denies that the effects follow necessarily from their causes and relates them to the will of their Creator, the will having no specific designated course but [a course that] can vary and change in kind, then let each of us allow the possibility of there being in front of him ferocious beasts, raging fires, high mountains, or enemies ready with their weapons [to kill him],but [also the possibility] that he does not see them because God does not create for him [Vision of them]. And if someone leaves a book in the house, let him allow as possible its change on his returning home into a beardless slave boy — intelligent, busy with his tasks — or into an animal; or if he leaves a boy in his house, let him allow the possibility of his changing into a dog; or [again] if he leaves ashes, [let him allow] the possibility of its change into musk; and let him allow the possibility of stone changing into gold and gold into stone. If asked about any of this, he ought to say: "I do not know what is at the house at present. All I know is that I have left a book in the house, which is perhaps now a horse that has defiled the library with its urine and its dung, and that I have left in the house a jar of water, which may well have turned into an apple tree. For God is capable of everything, and it is not necessary for the horse

> to be created from the sperm nor the tree to be created from the seed — indeed, it is not necessary for either of the two to be created from anything. Perhaps [God] has created things that did not exist previously. Indeed, if [such a person] looks at a human being he has seen only now and is asked whether such a human is a creature that was born, let him hesitate and let him say that it is not impossible that some fruit in the marketplace has changed into a human — namely, this human — for God has power over every possible thing, and this thing is possible; hence, one must hesitate in [this matter]. This is a mode wide open in scope for [numerous] illustrations, but this much is sufficient.
>
> (14) [Our] answer [to this] is to say:
>
> (15) If it is established that the possible is such that there cannot be created for man knowledge of its nonbeing, these impossibilities would necessarily follow. We are not, however, rendered skeptical by the illustrations you have given because God created for us the knowledge that He did not enact these possibilities. We did not claim that these things are necessary. On the contrary, they are possibilities that may or may not occur. But the continuous habit of their occurrence repeatedly, one time after another, fixes unshakably in our minds the belief in their occurrence according to past habit.

According to Gazali's thinking, not only are natural events such as the rising and the setting of the sun caused by God's direct intervention but when a supposed effect follows a supposed cause, such as when fire and cotton are placed in contact, the cotton is burned not because of the heat of the fire, but through God's direct intervention. However, because Gazali did not believe in a sun god, perhaps we can allow that he had no reason to suspect that the sun would do something different tomorrow. Verses 21. 33: " And He it is Who created the night and the day and the sun and the moon; all (orbs) travel along swiftly in their celestial spheres." and 36.40: "Neither is it allowable to the sun that it should overtake the moon, nor can the night outstrip the day; and all float on in a sphere." of the Koran seem to endorse both ancient Greek astronomy and a lawful regulation of the solar system.

The ancient Jews had a better understanding of the difference between intervention and operation-and-maintenance laws than modern science. When Isaiah told King Hezekiah that God was going to heal him, Hezekiah wanted a sign from God that Isaiah's message was from God. According to 2 Kings 20:8-11:

> And Hezekiah said unto Isaiah, What [shall be] the sign that the LORD will heal me, and that I shall go up into the house of the LORD the third day?
>
> And Isaiah said, This sign shalt thou have of the LORD, that the LORD will do the thing that he hath spoken: shall the shadow go forward ten degrees, or go back ten degrees? And Hezekiah answered, It is a light thing for the shadow to go down ten degrees: nay, but let the shadow return backward ten degrees.
>
> And Isaiah the prophet cried unto the LORD: and he brought the shadow ten degrees backward, by which it had gone down in the dial of Ahaz.

Instead of saying it as light thing for the shadow to speed up Hezekiah could have said it is almost a natural thing. That is, even if Hezekiah thought the rising and setting of the sun was due to God, it seems that he did not regard the rising and setting of the sun due to the intervention by God.

Gideon of the book of Judges also sought a sign from God to tell him whether or not God's hand was going to be with him to bring deliverance to Israel. According to Judges 6:36:

> And Gideon said unto God, If thou wilt save Israel by mine hand, as thou hast said, Behold, I will put a fleece of wool

> in the floor; [and] if the dew be on the fleece only, and [it be] dry upon all the earth [beside], then shall I know that thou wilt save Israel by mine hand, as thou hast said.
>
> And it was so: for he rose up early on the morrow, and thrust the fleece together, and wringed the dew out of the fleece, a bowl full of water.
>
> And Gideon said unto God, Let not thine anger be hot against me, and I will speak but this once: let me prove, I pray thee, but this once with the fleece; let it now be dry only upon the fleece, and upon all the ground let there be dew.
>
> And God did so that night: for it was dry upon the fleece only, and there was dew on all the ground.

Gideon was correct to question whether these effects were God's natural cause-and-effects or intervention effects and to request a second test. Gideon presumably knew enough to know that that he should put the fleece outside his tent and not under the shade of a tree, but until he did the first test, he possibly had not noticed that dew does not form on certain things. We should expect to get dew on grass exposed to a clear night sky, but not on rocks, and not on anything under the shade of a tree.

The ancient Jews had the understanding that there was one God — when God was present or when God was invoked then anything could happen but in other circumstances everything continued as normal. According to 1 Kings 20:23, following the defeat of the Syrians by the Jews: " And the servants of the king of Syria said unto him, Their gods are gods of the hills; therefore they were stronger than we; but let us fight against them in the plain, and surely we shall be stronger than they." Gentile cultures had lots of gods and perhaps more scope for ad hoc cause-and-effect explanations using their gods.

Gazali should have had a better understanding of the difference between naturalism and intervention, given that he only believed in one God. It could be that Gazali's 'occasionalism' was a religious reaction to the fact that philosophers wanted to explain too much in terms of natural laws. Gazali's error helps us to understand the error of modern science. Gazali asserted that operation-and-maintenance of the universe was carried out by God's intervention, whereas modern science asserts that God's creation is the result of the operation-and-maintenance laws. Gazali did not believe in operation-and-maintenance laws, whereas modern science does not believe in intervention.

10. CREATIONIST ARGUMENTS IN HOLBACH'S *SYSTEM OF NATURE*

Holbach's *System of Nature* provides some insight into early creationist arguments:

> [Part 1 Chapter V] THE observation of the necessary, regular, and periodical motion in the universe, generated in the mind of man the idea of ORDER; this term, in its original signification, represents nothing more than a mode of considering, a facility of perceiving, together and separately, the different relations of a whole; in which is discovered, by its manner of existing and acting, a certain affinity or conformity with his own. Man, in extending this idea to the universe, carried with him those methods of considering things which are peculiar to himself: he has consequently supposed there really existed in Nature affinities and relations, which he classed under the name of ORDER; and others which appeared to him not to conform to those, which he has ranked under the term of CONFUSION...
>
> Order, then, is nothing but necessity, considered relatively to the series of actions, or the connected chain of causes and effects, that it produces in the universe. What is the motion in our planetary system; but a series

of phenomena, operated upon according to necessary laws, that regulate the bodies of which it is composed? In conformity to these laws, the sun occupies the centre; the planets gravitate towards it, and revolve round it, in regulated periods: the satellites of these planets gravitate towards those which are in the centre of their sphere of action, and describe round them their periodical route. One of these planets, the earth which man inhabits, turns on its own axis; and by the various aspects which its revolution obliges it to present to the sun, experiences those regular variations which are called SEASONS. By a sequence of the sun's action upon different parts of this globe, all its productions undergo vicissitudes: plants, animals, men, are in a sort of morbid drowsiness during Winter: in Spring, these beings re-animate, to come as it were out of a long lethargy. In short, the mode in which the earth receives the sun's beams, has an influence on all its productions; these rays, when darted obliquely, do not act in the same manner as when they fall perpendicularly; their periodical absence, caused by the revolution of this sphere on itself, produces night and day. However, in all this, man never witnesses more than necessary effects, flowing from the nature of things, which, whilst that remains the same, can never be opposed with propriety. These effects are owing to gravitation, attraction, centrifugal power, &c...

It will no doubt be argued, that as Nature contains and produces intelligent beings, either she must be herself intelligent, or else she must be governed by an intelligent cause. We reply, intelligence is a faculty peculiar to organized beings, that it is to say, to beings constituted and combined after a determinate manner; from whence results certain modes of action, which are designated under various names; according to the different effects which these beings produce: wine has not the properties called wit and courage; nevertheless, it is sometimes seen that it communicates those qualities to men, who are supposed to be in themselves entirely devoid of them. It cannot be said Nature is intelligent after the manner of any of the beings she contains; but she can produce intelligent beings by assembling matter suitable to their particular organization, from whose peculiar modes of action will result the faculty called intelligence; who shall be capable of producing certain effects which are the necessary consequence of this property. I therefore repeat, that to have intelligence, designs and views, it is requisite to have ideas; to the production of ideas, organs or senses are necessary: this is what is neither said of Nature nor of the causes he has supposed to preside over her actions. In short experience warrants the assertion, it does more, it proves beyond a doubt, that matter, which is regarded as inert and dead, assumes sensible action, intelligence, and life, when it is combined and organized after particular modes.

[Part 2 Chapter V] It will be insisted that if a statue or a watch were shown to a savage, who had never before seen either, he would not be able to prevent himself from acknowledging that these things were the works of some intelligent agent of greater ability, possessing more industry than himself: it will be concluded from thence, that we are in like manner obliged to acknowledge that the universe, that man, that the various phenomena, are the works of an agent, whose intelligence is more comprehensive, whose power far surpasses our own. Granted: who has ever doubted it? the proposition is selfevident; it cannot admit of even a cavil. Nevertheless we reply, in the first place, that it is not to be doubted that nature is extremely powerful; diligently industrious: we admire her activity every time we are surprised by the extent, every time we contemplate the variety, every time we behold those complicated effects which are displayed in her works; or whenever we take the pains to meditate upon them: nevertheless, she is not really more industrious in one of her works than she is in another; she is not fathomed with more ease in those we call her most contemptible productions, than she is in her most sublime efforts: we no more understand how she has been capable of producing a stone or a metal, than the means by which she organized a head like that of the illustrious Newton. We call that man industrious who can accomplish things which we cannot; nature is competent to every thing: as soon therefore as a thing exists, it is a proof she has been capable of producing it: but it is never more than relatively to ourselves that we judge beings to be industrious: we then compare them to ourselves; and as we enjoy a quality which we call intelligence, by the assistance of which we accomplish things, by which we display our diligence, we naturally conclude from it, that those works which most astonish us, do not belong to her, but are to be ascribed to an intelligent being like ourselves, but in whom we make the intelligence commensurate with the astonishment these phenomena excite in us; that is to say, in other words, to our own peculiar ignorance, and the weakness incident to our nature.

In the second place, we must observe, that the savage, to whom either the statue or the watch is brought, will or will not have ideas of human industry: if he has ideas of it, he will feel that this watch or this statue, may be the work of a being of his own species, enjoying faculties of which he is himself deficient: if he has no idea of it, if he has no comprehension of the resources of human art, when he beholds the spontaneous motion of the watch, he will he impressed with the belief that it is an animal, which cannot be the work of man. Multiplied experience confirms this mode of thinking which is ascribed to the savage. The

Peruvians mistook the Spaniards for gods, because they made use of gunpowder, rode on horseback, and came in vessels which sailed quite alone. The inhabitants of the island of Tenian being ignorant of fire before the arrival of Europeans, the first time they saw it, conceived it to be an animal who devoured the wood. Thus it is, that the savage, in the same manner as many great and learned men, who believe themselves much more acute, will attribute the strange effects that strike his organs, to a genius or to a spirit; that is to say, to an unknown power; to whom he will ascribe capabilities of which he believes the beings of his own species are entirely destitute: by this he will prove nothing, except that he is himself ignorant of what man is capable of producing. It is thus that a raw unpolished people raise their eyes to heaven, every time they witness some unusual phenomenon. It is thus that the people denominate all those strange effects, with the natural causes of which they are ignorant, miraculous, supernatural, divine; but these are not by reasonable persons therefore considered proofs of what they assert: as the multitude are generally unacquainted with the cause of any thing, every object becomes a miracle in their eyes; at least they imagine God is the immediate cause of the good they enjoy--of the evil they suffer. In short, it is thus that the theologians themselves solve every difficulty that starts in their road; they ascribe to God all those phenomena, of the causes of which either they are themselves ignorant, or else unwilling that man should be acquainted with the source.

In the *third place*, the savage, in opening the watch, and examining its parts, will perhaps feel, that this machinery announces a work which can only be the result of human labour. He will perhaps perceive, that they very obviously differ from the immediate productions of nature, whom he has not observed to produce wheels made of polished metal. He will further notice, perhaps, that these parts when separated, no longer act as they did when they were combined; that the motion he so much admired, ceases when their union is broken. After these observations, he will attribute the watch to the ingenuity of man; that is to say, to a being like himself, of whom he has some ideas, but whom he judges capable to construct machines to which he is himself utterly incompetent. In short, he will ascribe the honour of his watch to a being known to him in some respects, provided with faculties very far superior to his own; but he will be at an immense distance from the belief, that this material work, whose ingenuity pleases him so much, can be the effect of an immaterial cause; or of an agent destitute of organs, without extent; whose action upon material beings cannot be within, the sphere of his comprehension. Nevertheless, man, when he cannot embrace the causes of things, does not scruple to insist that they are impossible to be the production of nature, although he is entirely ignorant how far the powers of this nature extend; to what her capabilities are equal. In viewing the world, we must acknowledge material causes for many of those phenomena which take place in it; those who study nature are continually adding fresh discoveries to this list of physical causes; science, as she enriches the intellectual stores of human enjoyment, every day throws a broader light on the energies of nature, which *prejudice*, aided by its almost inseparable companion, *ignorance*, would for ever bind down in the fetters of impotence.

Let us not, however, he told, that pursuing this hypothesis, we attribute every thing to a blind cause--to the fortuitous concurrence of atoms--to chance. Those only are called blind causes of which we know not either the combination, the laws, or the power. Those effects are called fortuitous, with whose causes man is unacquainted; to which his experience affords him no clue; which his ignorance prevents him from foreseeing. All those effects, of which he does not see the necessary connection with their causes, he attributes to chance. Nature is not a blind cause; she never acts by chance; nothing that she does would ever be considered fortuitous, by him who should understand her mode of action--who had a knowledge of her resources--who was intelligent in her ways. Every thing that she produces is strictly necessary--is never more than a consequence of her eternal, immutable laws; all is connected in her by invisible bonds; every effect we witness flows necessarily from its cause, whether we are in a condition to fathom it, or whether we are obliged to let it remain hidden from our view. It is very possible thereshould be ignorance on our part; but the words spirit, intelligence, will not remedy this ignorance; they will rather redouble it, by arresting our research; by preventing us from conquering those impediments which obstruct us in probing the natural causes of the effects, with which our visual faculties bring us acquainted.

This may serve for an answer to the clamour of those who raise perpetual objections to the partizans of nature, by unceasingly accusing them with attributing every thing to chance. Chance is a word devoid of sense, which furnishes no substantive idea; at least it indicates only the ignorance of its employers. Nevertheless, we are triumphantly told, it is reiterated continually, that a regular work cannot be ascribed to the concurrence of chance. Never, we are informed, will it be possible to arrive at the formation of a poem such as the Iliad, by means of letters thrown together promiscuously or combined at random. We agree to it without hesitation; but, ingenuously, are the letters which compose a poem thrown with the hand in the manner of dice? It would avail as much to say, we could not pronounce a discourse with the feet.

It is nature, who combines according to necessary laws, under given circumstances, a head organized in a mode suitable to bring forth a poem: it is nature who assembles the elements, which furnish man with a brain competent to give birth to such a work: it is nature, who, through the medium of the imagination, by means of the passions, in consequence of the temperament which she bestows upon man, capacitates him to produce such a masterpiece of fancy; such a never-fading effort of the mind: it is his brain modified in a certain manner, crowded with ideas, decorated with images, made fruitful by circumstances, that alone can become the matrix in which a poem can be conceived--in which the matter of it can be digested: this is the only womb whose activity could usher to an admiring world, the sublime stanzas which develop the story of the unfortunate Priam, and immortalize their author. A head organized like that of Homer, furnished with the same vigour, glowing with the same vivid imagination, enriched with the same erudition, placed under the same circumstances, would necessarily, and not by chance, produce the poem of the Iliad; at least, unless it be denied that causes similar in every thing must produce effects perfectly identical. We should without doubt be surprised, if there were in a dice-box a hundred thousand dice, to see a hundred thousand sixes follow in succession; but if these dice were all cogged or loaded, our surprise would cease: the particles of matter may be compared to cogged dice, that is to say, always producing certain determinate effects under certain given circumstances; these particles being essentially varied in themselves, countless in their combinations, they are cogged in myriads of different modes. The head of Homer, or of Virgil, was no more than an assemblage of particles, possessing peculiar properties; or if they will, of dice cogged by nature; that is to say, of beings so combined, of matter so wrought, as to produce

> the beautiful poems of the Iliad or the Aeneid. As much way be said of all other productions: indeed, what are men themselves but cogged dice-- machines into which nature has infused the bias requisite to produce effects of a certain description? A man of genius produces a good work, in the same manner as a tree of a good species, placed in a prolific soil, cultivated with care, grafted with judgment, produces excellent fruit.

Like Darwin and many modern day scientists, Holbach is overly impressed by the fact that the universe is operated-and-maintained by laws. Holbach effectively convinces himself that, because there are laws that make apples fall to the ground, there must be laws that create the apples. Like Thomas Hobbes, Holbach understands the effect of 'familiarity' on primitive people but he has no understanding of the effect of 'familiarity' (with say the fact that wheat grain can grow into wheat) on people who do not retain God in their thinking.

BIBLIOGRAPHY

- Abruzzo, Anthony J. *The Origins of the Nebular Hypothesis* (15 June, 2009) online @ The General Science Journal
- Aristotle: *Prior Analytics, Posterior Analytics, On Interpretation, Physics* online @ the University of Adelaide Library
- Bacon, Sir Francis: *Novum Organun* (1620) online @ the Online Library of Liberty
- Behe, Michael J.: *Darwin's Black Box --- The Biochemical Challenge to Evolution.* (1996) Touchstone -- Simon & Schuster
- Cairns-Smith A. G: *Seven Clues to the Origin of Life.* (1985) Cambridge University Press
- Creation Ministries International articles by various authors, *Arguments we think creationists should NOT use, Intelligent Design — 'A War on Science' says the BBC* online at Creation Ministries International.
- Dake, F. J: *Dake's Annotated Reference Bible* (1963) Dake Bible Sales, Inc.
- Darwin, Charles: *The Origin of Species.* (1968) Penguin Classics. (Edited by J.W. Burrow)
- Darwin, Charles: *The Descent of Man* online @ the University of Adelaide Library
- Dawkins, Richard : *The Blind Watchmaker.* (1986) Penguin Science
- Dawkins, Richard: *The God Delusion.* (2006) Black Swan
- Demere, Thomas A. and Walsh, Steve: *Facts, faith, and fairness* online @ SkepticReport
- Dennett, Daniel C.: *Darwin's Dangerous Idea --- Evolution and the Meanings of Life.* (1995) Penguin Science
- Desmond, Adrian & Moore, James: *Darwin.* (1992) Penguin Books

- Galileo, Galilei: *Dialogue Concerning Two New Sciences* online @ the University of Virginia
- Geisler, Norman: *Miracles and Modern Scientific Thought*
- Ghazali: The Incoherence of the Philosophers (2000) Brigham Young University Press
- Gould, Stephen Jay: Rocks of Ages --- Science and Religion in the Fullness of Life
- Grant, Edward : *The Foundations of Modern Science in the Middle Ages -- their religious, institutional and intellectual contexts.* (2008) Cambridge
- Harris, Sam: *The End of Faith.* (2004) Norton
- Hitchens, Christopher: *God is not Great.* (2007) Twelve
- Hobbes, Thomas: *Leviathan* (1651) @ The University of Adelaide Library
- Holbach, Paul-Henri Dietrich (Baron) : *System Of Nature* Parts 1 and 2 available online @ the Free Thought Archives
- Hume, David : *An Enquiry Concerning Human Understanding* (1748) online @ the University of Adelaide Library
- Johnson, Phillip: *The Church of Darwin*(1999) online @ Access Research Network (ARN)
- Lindberg David C: *The Beginnings of Western Science.* (2007) The University of Chicago
- Mill, John Stewart: *System of Logic* (1843) online @ the Online Library of Liberty
- Mills, David : *Atheist Universe - The Thinking Person's Answer to Christian Fundamentalism.* (2006) Ulysses Press
- National Academy of Sciences : *Science and Creationism --- A view from the National Academy of Sciences* (1999) National Academy Press
- NCSE articles by various authors: *Creationism and Pseudomathematics, Darwin Prosecuted: Review of Johnson's Darwin on Trial, Keep Science Free from Creationism, My Favourite Pseudo-science, "Science and Religion", "Christian Scholarship" and "Theistic Science", Voicesfor Evolution, "Intelligent Design" Not Accepted by Most Scientists, What is "Intelligent Design" Creationism?*

- Newton, Isaac: *The Principia.* (1995) Prometheus Books (Andrew Motte's translation of *Philosophiae Naturalis Principia Mathematica*)
- Patzia, Michael: *Anaxagoras* (16 June,2007) online @ Internet Encyclopedia of Philosophy
- Popper, Karl: *The Logic of Scientific Discovery.* (2002) Routledge
- Popper, Karl: *Conjectures and Refutations --- The Growth of Scientific Knowledge.* (2002) Routledge
- Popper, Karl : *Popper Selections edited by David Miller* (1985) Princeton
- Russell, Bertrand: *History of Western Philosophy.* (2000) Routledge
- Russell, Bertrand: *Why I Am Not A Christian* online @ the Drew University
- Schafersman Steven D.:*Naturalism is Today An Essential Part of Science* online @ Steven Jay Gould Archive
- Shapin, Steven: *The Scientific Revolution* (1998) The University of Chicago Press
- Slowik, Edward, *Descartes' Physics*, (2013) Edward N. Zalta (Ed.), online @ The Stanford Encyclopedia of Philosophy
- Sober, Elliot: *Philosophy of Biology* (1993) Oxford University Press.
- Spinoza, Benedict de: *The Chief Works of , Volume 1 (Tractatus-Theological-Politicus, Tractus Politicus*) (1670) online @ the Online Library of Liberty
- Theopedia: *God of the Gaps* online @ Theopedia
- Westfall, Richard S.: *Never at Rest --- A Biography of Isaac Newton.* (1980) Cambridge University Press
- Wikipedia articles by various authors: *Logical positivism, Vienna Circle, Watchmaker analogy, Teleological argument, Peppered moth evolution*
- Winston, Kimberly: *'New Atheists' Emerge From 9/11* (26 August 2001), online @ Huffington Post

Made in the USA
Monee, IL
05 November 2026

67811379R00213